ALWAYS SOMETIMES OR NEVER

Investigative statements that promote mathematical thinking, reasoning and explaining.

Ruth Bull and Priya Shah

MATHEMATICAL ASSOCIATION

Mα

Supporting mathematics in education

Always Sometimes or Never
© 2024 The Mathematical Association

Dedications

To Clive, Rachel and Joshua,
Who are "Always there", "Sometimes helpful" and "Never wrong!"
RB

To my mum, Lata, my inspiration.
To my dad, Jayant, my hero who showed me how to believe.
PS

Acknowledgements

This book came to fruition through collaboration and a commitment to embedding deeper understanding of mathematics for all learners. The journey became possible from our own individual experiences in working with students and the support from many, including our colleagues in and out of the classroom.

Our gratitude goes to Peter Bennett for his photographic expertise, guidance and unwavering support with the book cover, to Jonathan Ball for granting us permission to use MathsBot.com for publishing and to David Poras for granting us permission to use the free virtual manipulatives at Polypad.org, part of Amplify.

We also express our thanks to the Mathematical Association for their communications, commitment, time, support and dedication throughout the publishing process and thereafter. In addition, we give thanks to Allan Duncan, Elizabeth Glaister, Ray Huntley and Tom Roper, for your time, expertise and eyes as you ran through the drafts with a fine tooth comb.

Finally, we would like to thank our families and friends for their support, encouragement, their patience and understanding. We too have grown through the #mathschats about the content in this book.

From two mathematics enthusiasts to you:
Here's to mathematical growth...

Contents

Chapter	Title		Page
	Introduction...		1
	How to use the statements..		1
	A word on Manipulatives..		3
	A word on Proof in the Primary Setting...		3
	A word on Multiplication and Division...		4
1	**Addition and Subtraction Statements**		**7**
	AS001	The sum of two odd numbers is an odd number..............................	9
	AS002	The difference between two odd numbers is an even number........	12
	AS003	The sum of two even numbers is an even number............................	14
	AS004	The difference between two even numbers is an even number......	16
	AS005	The sum of an odd number and an even number is an even number...	18
	AS006	The difference between an even number and an odd number is an odd number...	19
	AS007	The sum of three odd numbers is an odd number............................	21
	AS008	The sum of three even numbers is an even number.........................	22
	AS009	The sum of three consecutive numbers is a multiple of three.........	24
	AS010	The difference between two numbers is always smaller than the minuend...	27
	AS011	When two numbers are added their order does not matter as the answer will be the same...	29
	AS012	Adding 2 to an even number results in an even number..................	30
	AS013	Adding 3 to an odd number results in an odd number.....................	32
	AS014	Swapping the order of two different numbers when finding their difference will result in the same answer..	34
	AS015	The sum of two 2-digit numbers will be a 3-digit number................	36
	AS016	50 can be made by adding three odd numbers..................................	38
	AS017	Adding 9 to a number is the same as adding 10 and subtracting 1......	40
	AS018	Adding 10 to a number results in a number that is a multiple of 10.....	41
	AS019	The sum of two numbers is the same as the product of those numbers.	43
	AS020	The difference between a 2-digit number and its reverse is a multiple of 9..	46
	AS021	The sum of four even numbers is divisible by 4.................................	49
	AS022	The square of an even number is divisible by 4.................................	51
	AS023	The sum of two triangular numbers is a square number..................	54
	AS024	Square numbers can be made by adding three triangular numbers......	56
	AS025	The sum of adding 4 to a negative number is a positive number.....	58
2	**Multiplication and Division Statements**		**61**
	MD001	The number 20 can be arranged as an array in only two different ways...	63
	MD002	The product of an odd number and an even number is an odd number..	65
	MD003	The product of three numbers is the same as the sum of them......	66
	MD004	Doubling a number makes it bigger..	67
	MD005	Halving a 2-digit number will result in a single digit.........................	68
	MD006	All numbers that end in 0 are multiples of 20....................................	70

Chapter	Title		Page
2	Continued		
	MD007	Doubling a multiple of 5 will give a multiple of 10	72
	MD008	Numbers (integers) that end in a 4 are multiples of 4	74
	MD009	To divide a number by 10, just remove the last digit	76
	MD010	A number that is divisible by 3 is also divisible by 6	78
	MD011	Multiples of 8 are also multiples of 6	80
	MD012	When a whole number is multiplied by 9, the sum of the digits of the product is 9	82
	MD013	The product of the diagonals of any 2x2 square on a multiplication grid are always equal (e.g. 6x12 and 9x8)	83
	MD014	Dividing a number by 1 will result in 1	85
	MD015	When a whole number is divided by 5 the biggest remainder there can be is 4	87
	MD016	An even number divided by an odd number gives a quotient with no remainder	90
	MD017	When multiplying two whole numbers, the resulting product will be greater than either of the chosen numbers	92
	MD018	Halving a 3-digit number less than 200 gives a 2-digit answer	94
	MD019	Finding a third of a number is the same as dividing it by 3	97
	MD020	Finding a quarter of a number is the same as halving and halving again	99
	MD021	The square of a number is greater than the number doubled	103
	MD022	A square number always has an odd number of factors	106
	MD023	Multiples of 6 are one more and one less than prime numbers	108
	MD024	Dividing a number makes the quotient smaller than the dividend	110
	MD025	Any whole number can be multiplied by partitioning (e.g. 35x4 is 30x4 plus 5x4)	113
3	**Fractions, Decimals and Percentages Statements**		**117**
	FD001	The larger the denominator of a fraction, the larger the fraction	119
	FD002	When the denominator is twice the numerator the fraction is worth 0.5	121
	FD003	The numerator of a fraction is less than the denominator	122
	FD004	Finding a fifth of a quantity is the same as multiplying by ten and halving the result	124
	FD005	The sum of two or more fractions can be calculated by adding the numerators and the denominators independently	126
	FD006	Fractions with the same denominator can be ordered in size	128
	FD007	If the numerator and the denominator are the same, then it is equal to one whole	130
	FD008	Fractions are less than one whole	132
	FD009	Doubling both numerator and denominator creates a new fraction that is double in size to the original one	133
	FD010	Dividing a fraction by 2 is the same as multiplying it by $\frac{1}{2}$	136
	FD011	10% of a number is the same as the number multiplied by 0.1	140
	FD012	Finding 5% of a number is the same as dividing the number by 5	142
	FD013	30% of 80 is the same as 80% of 30	145

Chapter	Title		Page
3	Continued		
	FD014	Reducing a number by 10% followed by increasing it by 10% results in returning to the original number..	147
	FD015	A quarter ($\frac{1}{4}$) can be written as 1.4..	149
	FD016	When comparing decimals, the longer the decimal the bigger the number..	151
	FD017	Decimal numbers can be written as fractions...	154
	FD018	Fractions can be simplified..	157
4	**Geometry and Shape Statements**		**159**
	GS001	A triangle has three acute angles..	161
	GS002	An equilateral triangle has all sides the same length, but the angles can be different..	165
	GS003	A triangle has perpendicular sides..	167
	GS004	An equilateral triangle has one line of symmetry...	169
	GS005	A square is a rectangle...	170
	GS006	A kite is a rhombus..	173
	GS007	A square has four lines of symmetry..	174
	GS008	A trapezium has two sides of equal length..	176
	GS009	Any quadrilateral can be made from two triangles..	178
	GS010	The interior angles of a pentagon total 360 degrees...	180
	GS011	A pentagon has two right angles..	182
	GS012	Hexagons have sides that are equal in length..	184
	GS013	The circumference is approximately three times the diameter...................................	187
	GS014	If the area of a rectangle is 24 squared centimetres, then the side lengths are 4 cm and 6 cm..	188
	GS015	The perimeter of a rectangle is four times one of the sides..	190
	GS016	The area of a triangle is $\frac{1}{2}$ x base x height..	192
	GS017	Doubling the width of a rectangle will result in the area being doubled...	195
	GS018	Two rectangles have the same perimeter, so they will have the same area..	197
	GS019	A pyramid has an even number of faces...	199
	GS020	A prism has at least three rectangular faces..	200
	GS021	A regular polygon will tessellate with itself..	202

Introduction

The aim of this book is to provide a bank of statements that can be used to promote deep mathematical thinking through an investigative approach.

The statements can be used by a student working with an adult, pair or group of students who will need to think and reason mathematically in an attempt to convince themselves and others about their findings. The use of manipulatives, diagrams, jottings or a combination of these should be used to investigate each statement. These should provide students with the support to explain their thoughts rather than just guessing.

We hope that by including 'answers' and guidance with each statement and by showing jottings with the manipulatives that this book can be used as a supportive tool to encourage mathematical discussions rich in reasoning and mathematical thinking. There are almost certainly other jottings or manipulatives that could also be used other than the ones provided in this book. Is there a more efficient manipulative and why? Along with the answers, you will find a section titled notes. Here we have provided supplementary information including possible connections to other statements or further investigations.

Each statement is a standalone activity and can be carried out independently of others. However, some explorations lead to other statements, in which case, references are made to the connecting statement numbers.

Digital manipulatives are used to produce this book, however, we believe there is great value in the sensory experiences with the manipulatives. With carefully crafted questions and supportive language used with the manipulatives, the mathematical concepts will come to life.

Maths by its very nature is interconnected, however for the purpose of this book, the statements have been placed into 4 categories: Addition and Subtraction; Multiplication and Division; Fractions, Decimals and Percentages; Geometry and Shape.

How to use the statements

One way to use the statements is to choose a statement and give it to a pair or group of students to discuss and investigate, whilst providing support through careful questioning and promoting the use of correct mathematical language alongside the use of manipulatives.

Many of the statements have been tried and tested and one approach that has worked effectively is to provide the statement to each group of students on a small piece of paper, which can be either copied onto a bigger piece of paper (A4 or even A3) or stuck in the centre of the larger paper. Students can place their jottings, examples and thoughts onto the paper - around the statement before eventually arriving to a conclusion. The words 'Always', 'Sometimes' and 'Never' should be written at the beginning of the task on the side of the paper, so that a line can be drawn to indicate their conclusion. Two different examples are shown on the following page.

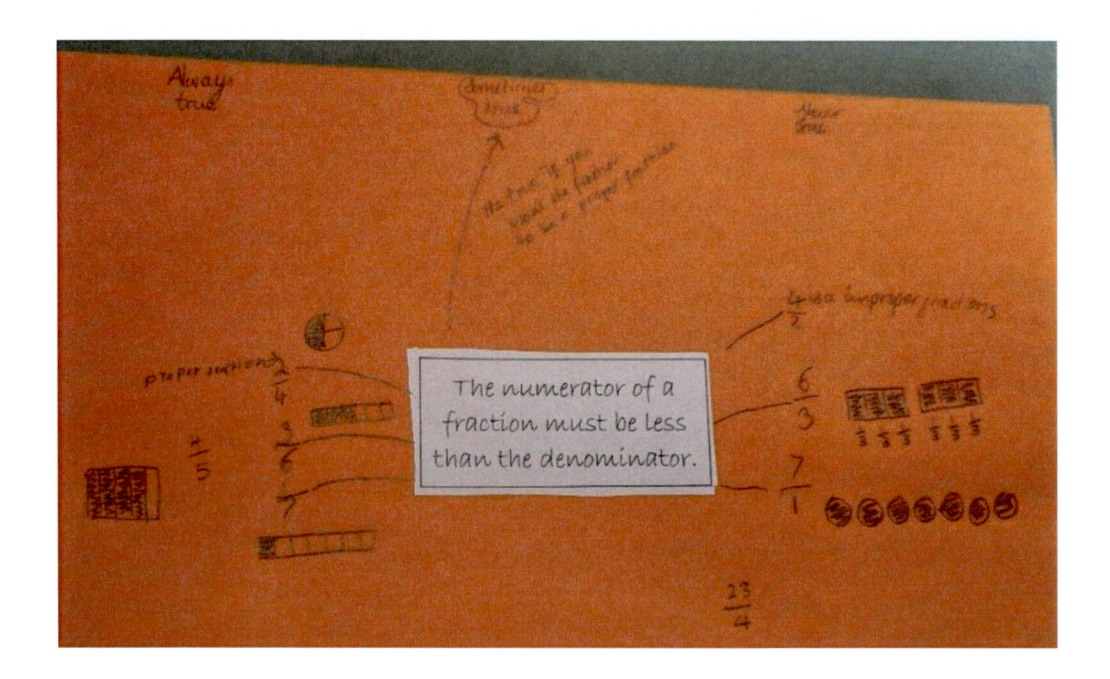

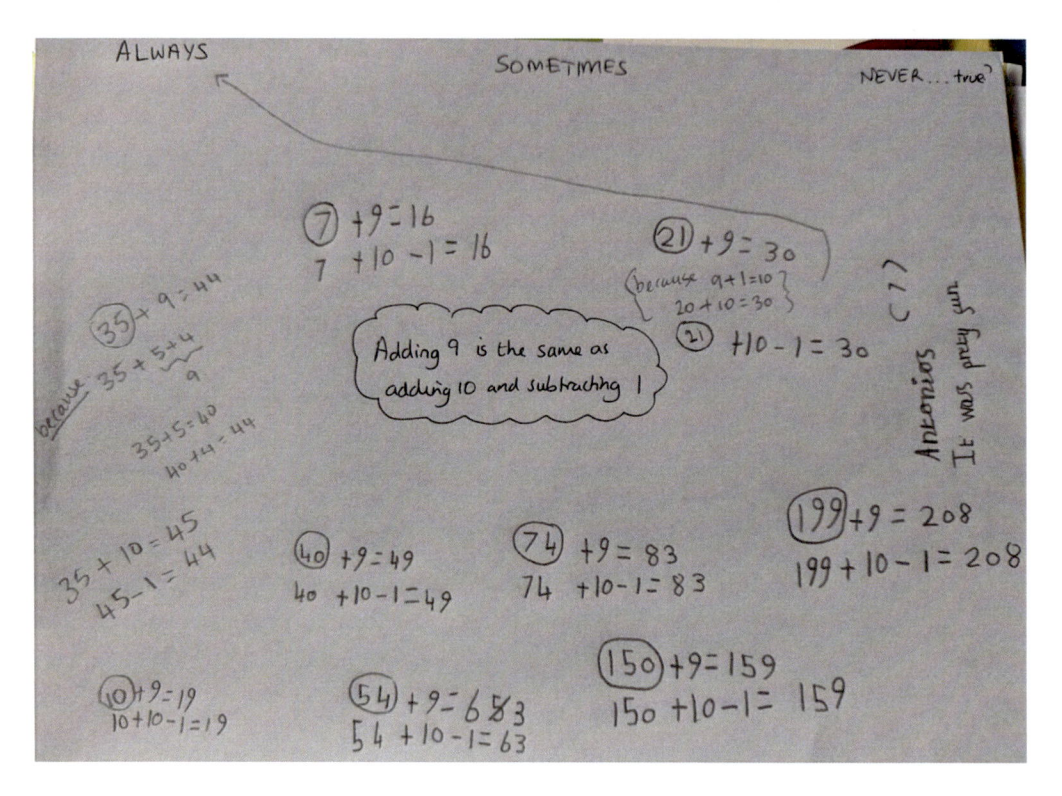

The same approach can be used whilst working on a one-to-one basis with a student or as a whole class collaborative activity.

You will notice that no time frame per activity has been provided in the supplementary notes as we found that to fulfil the purpose of the activity, the time varied per student, per group of students and per class. Therefore, the time allocation judgement lies with you, the educator.

For ease of printing, the collated statements per category can be found at the beginning of each chapter.

A word on Manipulatives

Many different manipulatives are used in this book. Some may use different names to those you are familiar with, for example: 'Base 10 apparatus' instead of Dienes. If you are unable to access the manipulative mentioned in a statement, then consider other alternatives. Such as, using strips of paper as a substitute for fraction bars. Some types of household packaging are useful as 3D shape models, for example cereal boxes as cuboids, pyramid shaped teabags and cube shaped tissue boxes.

A few words about Cuisenaire rods: The beauty about them is that they are proportionate in size, thus accommodating for any value to be assigned to one of them. For example, if the white unit rod is assigned the value of 1, then the orange rod is assigned the value of 10, as seen in the diagram (created using digital manipulatives from MathsBot.com).

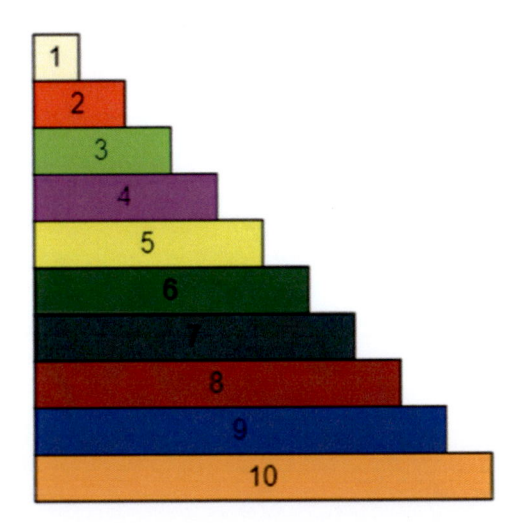

A word on Proof in the Primary Setting

Mathematical proof is generally identified as a formal and rigorous explanation using algebraic notation. In most cases, in schooling, it is used to produce a generalisation. We feel that the idea of generalisation can be employed early in mathematics learning and with this in mind, we approach it as a route towards generalisations via the act of convincing, communicating, investigating and justifying.

The aim is to encourage students to try examples and think deeply about the statement they have chosen or assigned and to test out their ideas using different conditions. Hence, trying out a particular case and then, investigate further by slightly changing one of the conditions – Would the outcome remain the same? Would it be different? Why?

For example, trying a statement that involves an odd number and an even number, then considering if the outcome would be the same if the chosen numbers were both even numbers or if they were both odd numbers. With exposure and experience of these type of activities, students behave as mathematical thinkers, mathematical communicators and mathematical problem solvers.

In secondary and higher education, proofs are written through a variety of arguments, posed conditions, algebraic notations and much more. A taste of this will be seen in some of the chapters with the use of algebra and also 'proof by contradiction' (used in one of the statements).

This book can be seen as a stepping stone towards mathematical proofs.

A word on Multiplication and Division

We appreciate that different explanations about multiplication exist around the world. As a result, we explain our interpretations below along with its respective inverse – division.

Some terms used with multiplication and division in equation form are:

1) multiplicand multiplied by multiplier is equal to product
2) factor multiplied by factor is equal to product
3) dividend divided by divisor is equal to quotient

How these are written in multiplication and division number sentences, read, interpreted, translated and represented are explained on the following page through examples using Cuisenaire rods and counters from MathsBot.com.

Multiplication Sentence	Read as	Interpreted as	Representation of the interpretation
5 x 4 = 20	five multiplied by four is equal to twenty	five multiplied four times	5 / 5 / 5 / 5
5 x 4 = 20	five times four is equal to twenty	five groups of four or five lots of four	4 4 4 4 4
5 x 4 = 20	five multiplied by four is equal to twenty	five multiplied four times	(4 rows of 5 circles)
5 x 4 = 20	five times four is equal to twenty	five groups of four or five lots of four	(5 columns of 4 circles)

Division Sentence	Read as	Interpreted as	Representation of the interpretation
20 ÷ 5 = 4	twenty divided by five is equal to four	How many fives divide into twenty? or How many groups of five are there in twenty? This interpretation is known as quotative division, where the focus is on finding the number of groups.	
20 ÷ 5 = 4	twenty divided by five is equal to four	Share twenty equally between five groups. This interpretation is known as partitive division, where the focus is on sharing equally to determine the quantity in each group.	
20 ÷ 5 = 4	twenty divided by five is equal to four	How many fives divide into twenty? or How many groups of five are there in twenty? This interpretation is known as quotative division, where the focus is on finding the number of groups.	
20 ÷ 5 = 4	twenty divided by five is equal to four	Share twenty equally between five groups. This interpretation is known as partitive division, where the focus is on sharing equally to determine the quantity in each group.	

Chapter 1

Addition and Subtraction Statements

The sum of two odd numbers is an odd number. AS001	The sum of three even numbers is an even number. AS008
The difference between two odd numbers is an even number. AS002	The sum of three consecutive numbers is a multiple of three. AS009
The sum of two even numbers is an even number. AS003	The difference between two numbers is always smaller than the minuend. AS010
The difference between two even numbers is an even number. AS004	When two numbers are added their order does not matter as the answer will be the same. AS011
The sum of an odd number and an even number is an even number. AS005	Adding 2 to an even number results in an even number. AS012
The difference between an even number and an odd number is an odd number. AS006	Adding 3 to an odd number results in an odd number. AS013
The sum of three odd numbers is an odd number. AS007	Swapping the order of two different numbers when finding their difference will result in the same answer. AS014

The sum of two 2-digit numbers will be a 3-digit number. AS015	The sum of four even numbers is divisible by 4. AS021
50 can be made by adding three odd numbers. AS016	The square of an even number is divisible by 4. AS022
Adding 9 to a number is the same as adding 10 and subtracting 1. AS017	The sum of two triangular numbers is a square number. AS023
Adding 10 to a number results in a number that is a multiple of 10. AS018	Square numbers can be made by adding three triangular numbers. AS024
The sum of two numbers is the same as the product of those numbers. AS019	The sum of adding 4 to a negative number is a positive number. AS025
The difference between a 2-digit number and its reverse is a multiple of 9. AS020	

> **AS001 Statement:** The sum of two odd numbers is an odd number.
>
> **Answer:** NEVER TRUE
>
> **Manipulatives:** Number Frames, Number Lines

Using Number Frames

This statement can be shown as never true through the use of *Number Frames* by physically joining 2 odd-number frames.

One can generalise that an odd number added to an odd number yields an even number. The evenness is shown through the physical equal pairing, along with number sentences to support the findings. Examples of the arrangement of number frames with different odd numbers are shown in the diagrams below through the use of odd numbers smaller than 10.

1 + 3 = 4

1 + 5 = 6

1 + 7 = 8

Examples of the arrangement of number frames with both odd numbers being the same are shown in the diagrams below, again through the use of odd numbers smaller than 10.

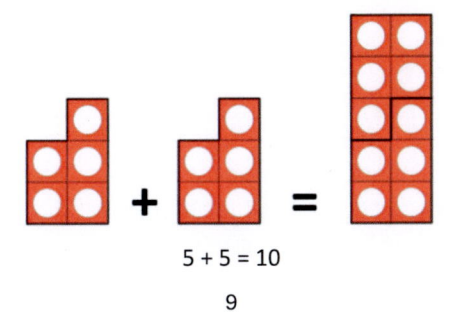

5 + 5 = 10

$7 + 7 = 14$

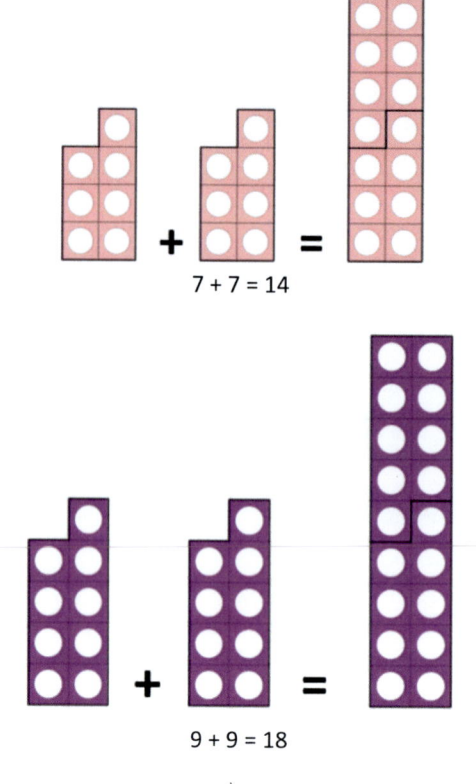

$9 + 9 = 18$

From the examples above, the statement is never true. Encourage students to explore the statement through several different combinations of odd numbers to lead them to generalise their findings.

Using Number Lines

This statement can be shown as never true through the use of *Number Lines* by showing the addition of two odd numbers as adding on/counting on. In the examples below, the same odd number has been used instead of two different odd numbers.

The fact that the sum is an even number can be identified by the recognition of the number on the number line as even. Alternatively, the number frames can be used alongside the number lines, showing the symmetry in the sum, declaring it to be an even number. Similarly, number sentences can be written to support the findings alongside the diagrams.

Examples of the number lines are shown in the diagrams below.

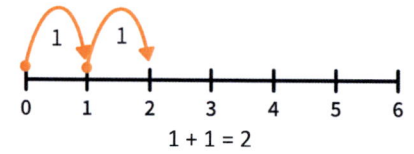

$1 + 1 = 2$

Students may show their understanding using the empty number line with counting on shown in the diagram on the following page.

$1 + 1 = 2$

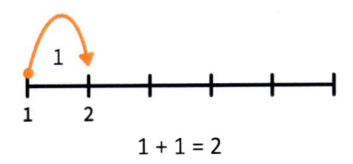

Another example:

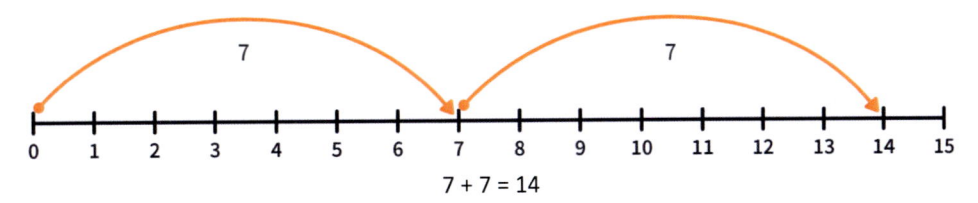

$7 + 7 = 14$

Students may show their understanding using number bonds with counting on as shown in the diagram below.

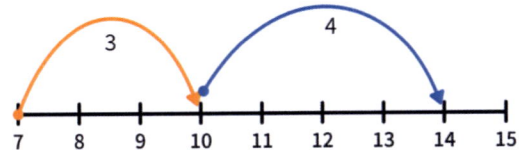

> **Notes:** Number Frames used in the diagrams are from MathsBot.com and the Number Lines in the diagrams are from Polypad.org.
>
> Students may use other orientations of the number frames instead of those shown above. Diagrams that resemble the number frames (dots in square paper) can be used as a substitute for the number frames.
>
> Students could investigate further by making a prediction for the statements: 'The sum of two even numbers is even' or 'The sum of an even number and an odd number is even/odd'. Thus, leading them towards generalisations with whole numbers. See AS003.
>
> All odd and even numbers should be whole numbers. Odd numbers have a ones (units) digit of either 1,3,5,7 or 9. Even numbers have a ones (units) digit of either 0,2,4,6 or 8.

> **AS002 Statement:** The difference between two odd numbers is an even number.
>
> **Answer:** ALWAYS TRUE
>
> **Manipulatives:** Number Frames

This statement can be shown as always true through the use of *Number Frames* by physically placing 2 odd-number frames on top of each other where the larger number is placed at the bottom. In this case, the difference is identified as the part of the larger number still visible after placing the smaller number on top.

The evenness can be seen through equal pairing or symmetry, including recognition through the value of the difference.

Examples of the arrangement of number frames is shown in the diagrams below through the use of odd numbers smaller than 10 alongside number sentences to support the findings.

Let's first consider using number frames to represent 9 – 7. Students may place the 7 number frame on top of the 9 number frame in different ways as seen below. The assumption here is that the number frame that is placed on top is in fact hiding the exact quantity that is below it, thus revealing the difference as the quantity seen in the same colour as the minuend. In this case, the value of the difference is 2.

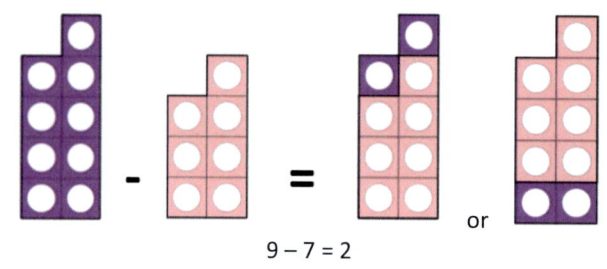

9 – 7 = 2

or

Students can use a piece of paper/card to hide the value of the subtrahend from the minuend to reveal the difference. The above example of 9 – 7 can be represented like this:

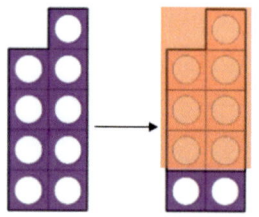

Now let's consider using number frames to represent 9 – 3.

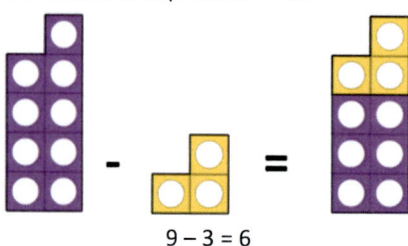

9 – 3 = 6

Again, students may place the 3 number frame on top of the 9 number frame in different ways as seen here: or or

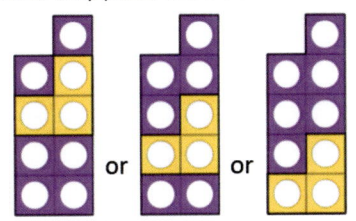

Irrespective of the arrangements, it can be seen that the statement is true for the examples above. Encourage students to explore with several odd numbers and articulate their findings.

> **Notes:** Number Frames used in the examples are from MathsBot.com.
>
> Students may use other orientations of the number frames instead of those shown above. Another form of representation can be to show by cancelling or removing the subtrahend from the minuend to reveal the value of the difference. However, in this case the evenness may not be obvious like some of the diagrams above without rearrangement or recognition of symmetry.
>
> Diagrams that resemble the number frames (dots in square paper) can be used as an alternative. Also, other manipulatives such as number lines and Dienes may be used. Be aware that the physical manipulation for some larger numbers may become cumbersome with Dienes.
>
> Further predictions such as 'The difference between two even numbers is an odd number', maybe made by the student. See AS004.
>
> All odd and even numbers should be whole numbers. Odd numbers have a ones (units) digit of either 1,3,5,7 or 9. Even numbers have a ones (units) digit of either 0,2,4,6 or 8.

> **AS003 Statement:** The sum of two even numbers is an even number.
>
> **Answer:** ALWAYS TRUE
>
> **Manipulatives:** Number Frames, Cuisenaire Rods

Using Number Frames

This statement can be shown as always true through the use of *Number Frames* by physically joining two even number frames. This can be achieved through joining the frames above/below each other or on the side. The evenness can be seen through equal pairing and the symmetry.

Examples of the use of number frames are shown in the diagrams below, , as well as, writing number sentences to support the findings.

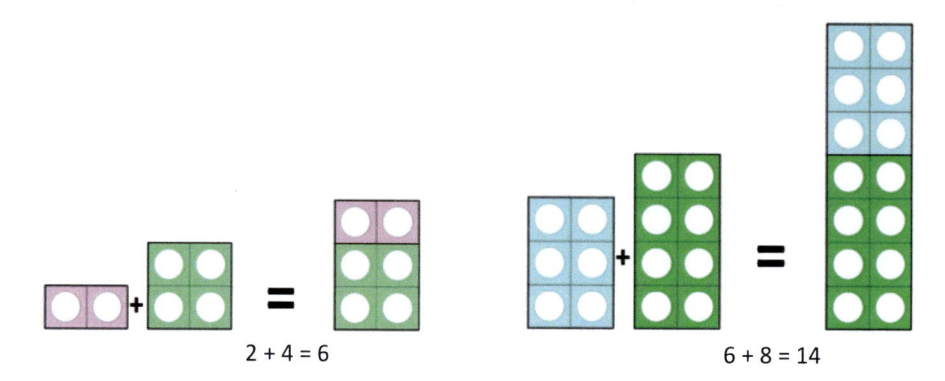

2 + 4 = 6   6 + 8 = 14

From the examples above, it can be seen that the statement is always true. Note that students should explore several different examples to generalise their findings. How could this be explored if one of the numbers is zero? If both numbers are zero?

Using Cuisenaire Rods

The same examples as above are shown using *Cuisenaire rods* in the diagrams below. In these diagrams, they are shown as physically adding horizontally to support the visual image of how numbers increase on a number line from left to right and the middle bar shows their sum. To further reveal the evenness, the equal valued bars are placed on the top. This lands itself neatly to connect with the concepts of doubling and halving too.

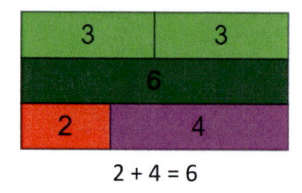

2 + 4 = 6

In the next diagram, as the sum is greater than 10, the number bars that support the efficient calculation are shown under the bracketed sum of 14.

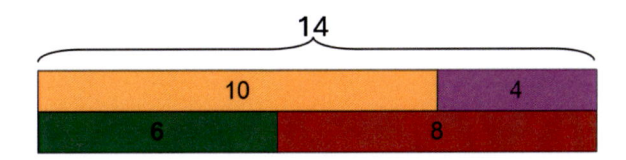

To further reveal the evenness, the equal valued bars are placed on the top (as seen in the diagram below). Again, creating the opportunity to discuss symmetry along with doubling and halving.

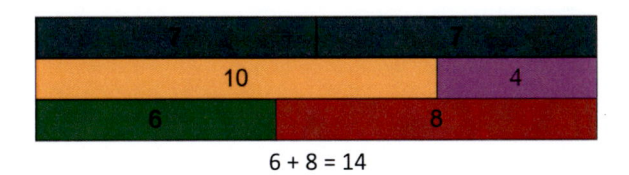

6 + 8 = 14

Again, it can be seen that the statement is always true.

> **Notes:** Number Frames and Cuisenaire Rods used in the examples are from MathsBot.com.
>
> Students may use other orientations of the number frames instead of those shown above. When using Cuisenaire rods as seen in the diagrams above, it gives an opportunity to discuss subtractions. Thus, supporting the connection between subtraction and addition as an inverse relationship – additive reasoning.
>
> Other manipulatives such as Dienes may be used for larger numbers. Be aware that the physical manipulation for some larger numbers may become cumbersome. Diagrams that resemble the number frames (dots in square paper) and rectangles that represent the Cuisenaire rods can be drawn as an alternative.
>
> Students may further predict what happens when more even numbers are added. Thus, providing further investigative opportunities.
>
> All even numbers should be whole numbers and have a ones (units) digit of either 0,2,4,6 or 8.

> **AS004 Statement:** The difference between two even numbers is an even number.
>
> **Answer:** ALWAYS TRUE
>
> **Manipulatives:** Number Frames, Place Value Counters

Using Number Frames

This statement can be shown as always true through the use of *Number Frames* by physically hiding an even number from the larger piece (minuend) to reveal the difference. Here the amount that is subtracted is shown by hiding using a piece of coloured paper/card. Note that the colour of the paper/card is not of relevance here, the act of hiding is. This is done to support the mental visualisation of both the difference between the numbers and the action of (taking away) subtraction.

Students should be able to show that the value of the difference can be identified as a multiple of 2 or is divisible by 2. This can be shown in calculations especially for large numbers without diagrams.

Examples of the arrangement of number frames is shown in the diagram below through the use of even numbers up to 20.

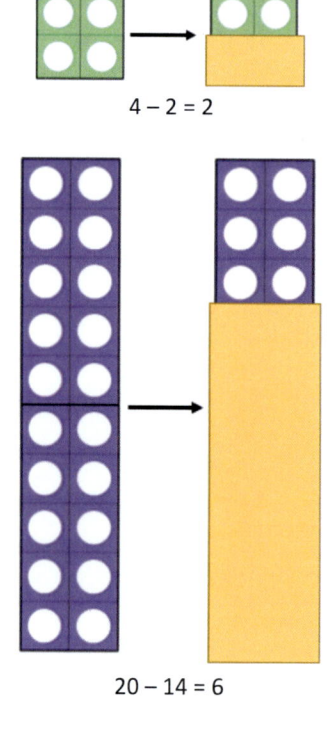

$4 - 2 = 2$

$20 - 14 = 6$

The above examples suggest the statement to be always true. Encourage students to show several examples with even numbers to support leading them to predict and generalise.

Using Place Value Counters

This statement can be shown as always true through the use of *Place Value Counters* by physically hiding or removing an even number value from the larger even number to reveal the difference. When using place value counters, the total value of the counters must be considered and identified as an even number from the beginning. For example, the diagram below shows the total value of 3330 but, uses an odd number of counters.

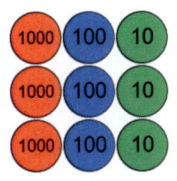

In the examples below, the act of subtraction is performed by crossing out the value of the subtrahend from the minuend to reveal the difference. Encourage the students to show that the value of the difference can be identified as a multiple of 2 or is divisible by 2.

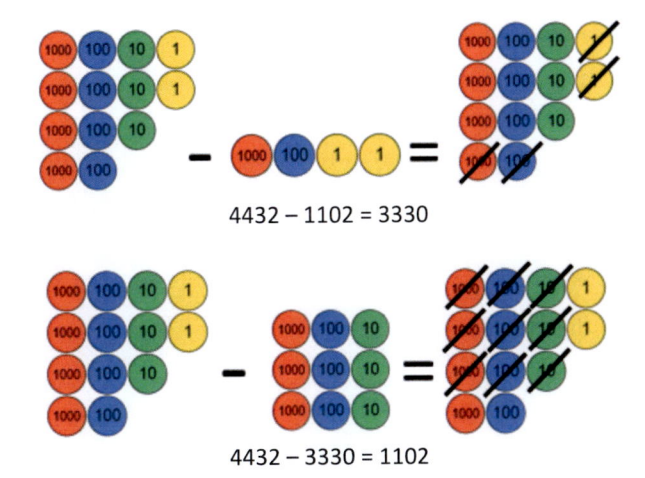

4432 – 1102 = 3330

4432 – 3330 = 1102

From all the examples above, it can be seen that the statement is always true.

> **Notes:** Number Frames and Place Value Counters used in the examples are from MathsBot.com.
>
> Students may use other arrangements of the number frames or counters instead of those shown above including other ways of hiding or eliminating the number that is subtracted, respectively.
>
> All even numbers should be whole numbers with the ones (units) digit of either 0,2,4,6 or 8.
>
> A prediction such as 'The difference between two odd numbers is an odd number', maybe made by the students thus, giving an opportunity to investigate further. See AS002.

> **AS005 Statement:** The sum of an odd number and an even number is an even number.
>
> **Answer:** NEVER TRUE
>
> **Manipulatives:** Number Frames

This statement can be shown as never true through the use of *Number Frames* by physically joining odd and even number frames as an act of addition. Note that the order of addition can be either odd number + even number or vice versa. This provides an opportunity to discuss the commutative property of addition.

Examples of arrangement of number frames shown in the diagrams below highlight the lack of evenness and the lack of symmetry in the arrangements of the sum. Note that the number frame of value 3 does have symmetry through a diagonal, which creates two halves. Therefore, clarity must be provided when discussing the symmetry of an even number by relating it to whole numbers.

Let us consider two examples with the addends to be less than 10.

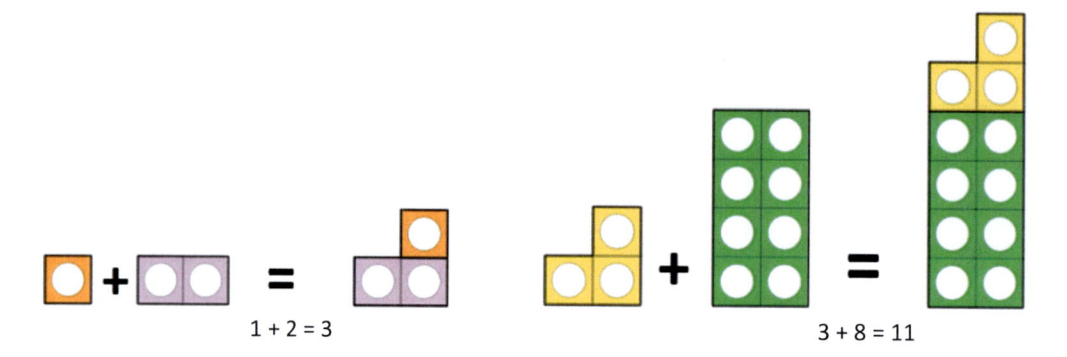

$1 + 2 = 3$ $3 + 8 = 11$

From the examples above, the statement appears to be never true. Encourage students to explore the statement with several examples to justify their findings. Can the student identify whether the value of the sum will be an odd number or an even number without the use of manipulatives?

> **Notes:** Number Frames used in the examples are from MathsBot.com.
>
> Students may use other orientations of the number frames instead of those shown above.
>
> Predictions about the sum of two even numbers and the sum of two odd numbers may be made by the student thus, giving an opportunity to investigate further. See AS001 and AS003.
>
> All odd and even numbers should be whole numbers. Odd numbers have a ones (units) digit of either 1,3,5,7 or 9. Even numbers have a ones (units) digit of either 0,2,4,6 or 8.

AS006 Statement: The difference between an even number and an odd number is an odd number.

Answer: ALWAYS TRUE

Manipulatives: Number Frames

This statement can be shown as always true through the use of *Number Frames* by physically comparing the even and odd number frames. This can be achieved through placing frames on top of each other or through crossing out/ hiding the smaller value from the larger value as a subtraction. The oddness can be seen through unequal pairing in the difference with the assumption that the frame is symmetrical if it has an even number of holes.

Diagrams that resemble the number frames (dots in square paper) and number sentences can be written to support the findings.

Examples of an even number minus an odd number is shown in the diagrams below, where both the crossing out and identification of the difference is made visible.

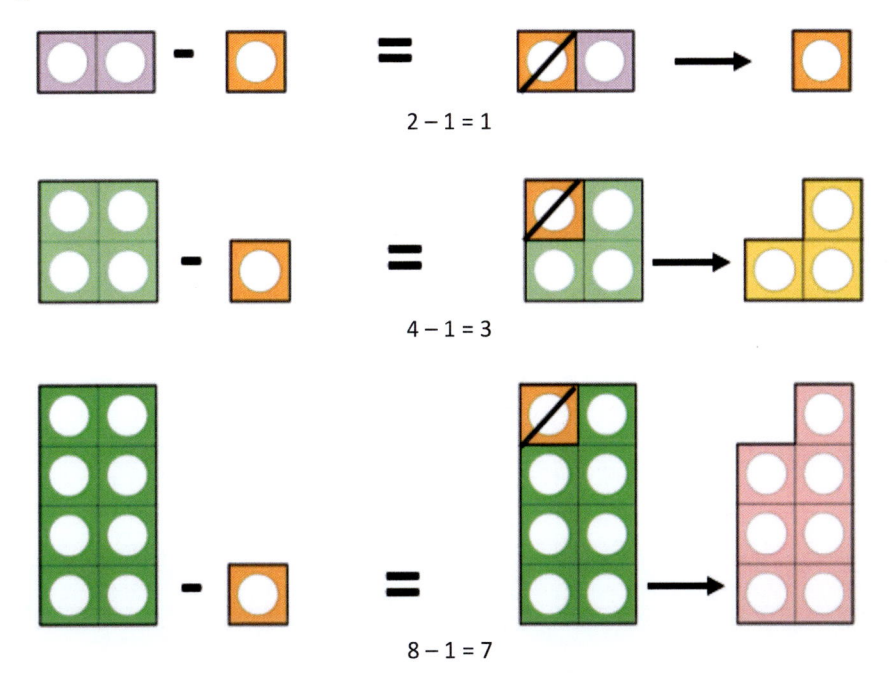

$2 - 1 = 1$

$4 - 1 = 3$

$8 - 1 = 7$

An example of an odd number minus an even number is shown in the examples on the following page. Again, both the crossing out and identification of the difference is explicitly shown. Notice, in this case, the odd and even numbers have been created using different number frames. The number 25 is represented by using a 10 frame, an 8 frame and a 7 frame and the number 18 is represented by using an 8 frame, a 6 frame and a 4 frame. This is intentional as it supports the composition of numbers in a variety of combinations. Encourage students to create their numbers using a variety of combinations. Does the composition of the number change the findings?

$$25 - 18 = 7$$

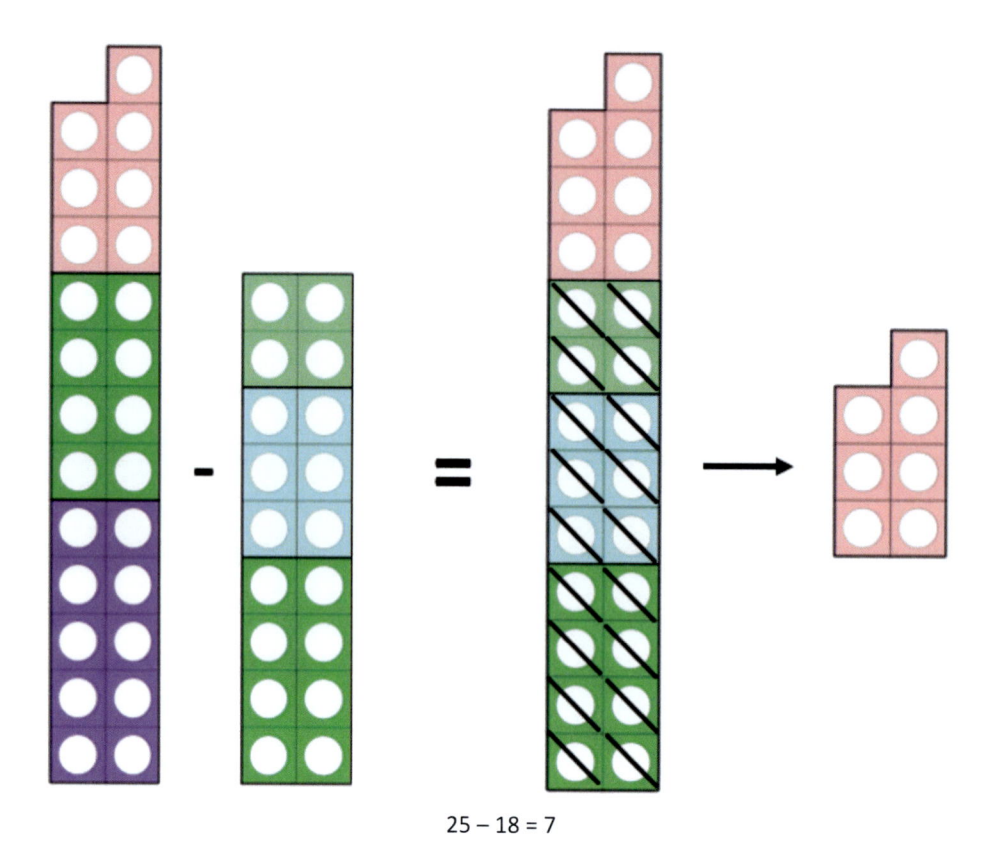

From the examples, the statement is always true. Therefore, students should be able to conclude that the difference is always an odd number irrespective of whether the minuend is an even number or an odd number.

> **Notes:** Number Frames used in these examples are from MathsBot.com.
>
> Students may use other orientations of the number frames instead of those shown above, including other combinations of number frames to create their minuend and subtrahend.
>
> Other manipulatives such as Dienes or maths link cubes may be used for larger numbers. Be aware that the physical manipulation for some larger numbers may become cumbersome.
>
> See AS002 and AS004.
>
> All even and odd numbers should be whole numbers. Even numbers have a ones (units) digit of either 0, 2, 4, 6 or 8 and odd numbers have a ones (units) digit of either 1, 3, 5, 7 or 9.

> **AS007 Statement:** The sum of three odd numbers is an odd number.
>
> **Answer:** ALWAYS TRUE
>
> **Manipulatives:** Number Frames

This statement can be shown as always true through the use of *Number Frames* by physically joining 3 odd number frames as an act of addition.

Examples of two different arrangements of the same sum is shown in the diagrams below through the use of odd numbers smaller than 10 alongside number sentences to support the findings.

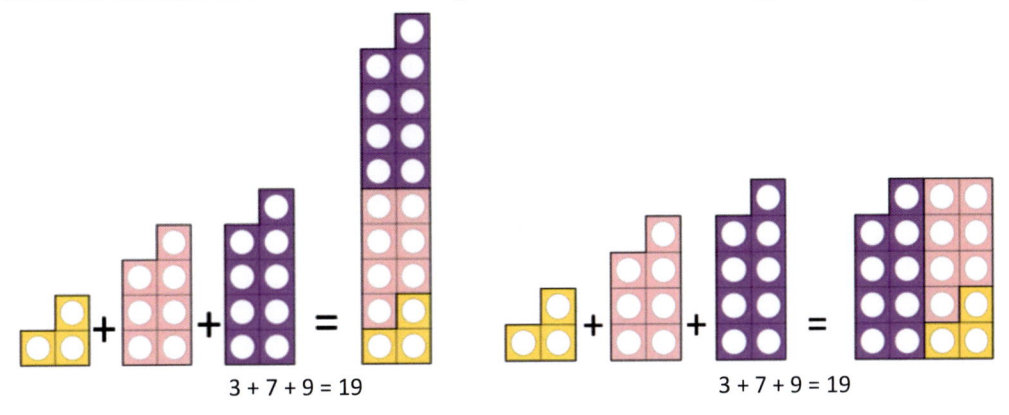

$3 + 7 + 9 = 19$ $3 + 7 + 9 = 19$

As students manipulate with their choice of numbers, encourage them to articulate their findings about the partial sum in the process – the sum of two odd numbers results in an even number.

Although the above example is only one case, and when investigating, we should explore several examples to conjecture and predict. However, from the previous work on AS001 and AS005, alongside this, we can state that the statement is always true.

Can students find a pattern in the evenness or oddness of the sum in relation to the number of odd numbers that are added?

> **Notes:** Number Frames used in the examples are from MathsBot.com.
>
> Students may use other orientations of the number frames instead of those shown above. Other manipulatives such as Dienes or maths link cubes may be used for larger numbers. Be aware that the physical manipulation for some larger numbers may become cumbersome.
>
> Further predictions such as 'The sum of three even numbers is an even number', maybe made by the students thus, giving an opportunity to investigate further.
>
> All odd numbers should be whole numbers. Odd numbers have a ones (units) digit of either 1,3,5,7 or 9.

AS008 Statement: The sum of three even numbers is an even number.

Answer: ALWAYS TRUE

Manipulatives: Number Frames

This statement can be shown as always true through the use of *Number Frames* by physically joining 3 even number 'number frames' as an act of addition. The evenness can be seen through equal pairing in the sum.

Example of an arrangement is shown in the diagram below through the use of even numbers up to 10 along with the number sentence. Diagrams that resemble the number frames (dots in square paper) can be used as an alternative.

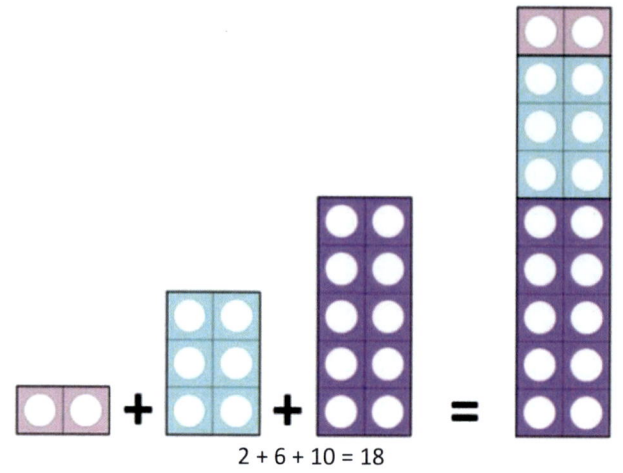

2 + 6 + 10 = 18

The sum can be arranged in different orientations as in the diagrams below. However, note that the symmetry is not obvious and so may result in the student concluding that the sum is not even. In such situations, encourage different arrangements so that the evenness is visually apparent.

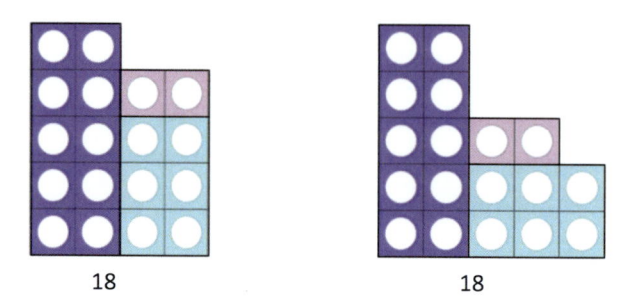

18   18

Students may use an alternative approach by considering their findings from the partial sum – the sum of two even numbers results in an even number. Then, using the value of the partial sum as the new addend (which is an even number) and add it to their third even number. In the case of the above example, 2 + 6 = 8 (seen as an even number plus an even number is equal to an even number) then, 8 + 10 = 18.

Although the above example is only one case, and when investigating, we should explore several examples to conjecture and predict. However, from the previous work on AS003, alongside this, we can state that the statement is always true.

Can students generalise that any number of even numbers will always result in their sum being an even number?

> **Notes:** Number Frames used in these images are from MathsBot.com.
>
> Students may use other orientations of the number frames instead of those shown above.
>
> Other manipulatives such as Dienes or maths link cubes may be used for larger numbers. Be aware that the physical manipulation for some larger numbers may become cumbersome.
>
> All even numbers should be whole numbers and have a ones (units) digit of either 0, 2, 4, 6 or 8.

> **AS009 Statement:** The sum of three consecutive numbers is a multiple of three.
>
> **Answer:** ALWAYS TRUE
>
> **Manipulatives:** Counters, Number Lines or Dienes

Using Counters

This statement can be shown as always true through the use of *Counters* by physically arranging them so that they form a rectangular structure such that the counters can be seen in groups of 3.

The use of the counters supports the visualisation of the 'groups of three' structure even if the multiple of 3 is unknown by the students. Diagrams that resemble the counters like dots or circles on square paper can be drawn instead, as well as, writing number sentences to support the findings.

Examples of the use of counters are shown in the diagrams below. Notice that the diagrams shown below show vertical grouping of 3 (3 rows). Students may use other orientations when using the counters. Ensure that their choice of display is vivid in the grouping of 3 to support their findings.

$$1 + 2 + 3 = 6$$

$$3 + 4 + 5 = 12$$

From these examples, the statement appears to be true.

Using Number Lines

The diagrams below show the jumps of consecutive numbers on a number line as cumulative. The multiples of 3 are marked by using coloured dots on their respective place on the number line. Even though this requires pre-mark making, it provides a confirmation of the sum being a multiple of 3.

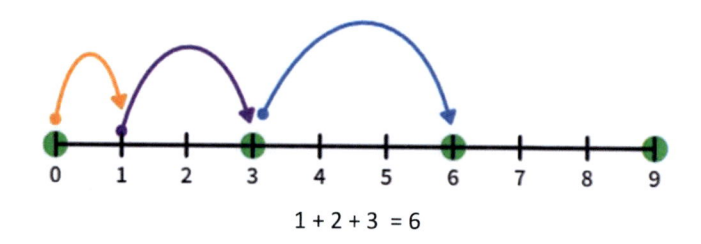

$$1 + 2 + 3 = 6$$

The diagram above shows the sum of adding 1, 2 and 3 respectively and the diagram below shows the sum of adding 2, 3 and 4 respectively.

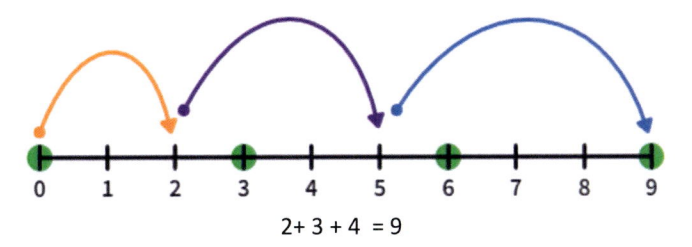

$$2 + 3 + 4 = 9$$

Again for these examples, the statement appears to be true.

Using Dienes

Example of the arrangement of three large consecutive numbers 99, 100 and 101 is as shown in the diagram below using Dienes blocks. Notice the rearrangement to support the regrouping and renaming, thus, resulting in the sum equating to the value of 300. The 3 groups of 100 confirms the sum being a multiple of 3.

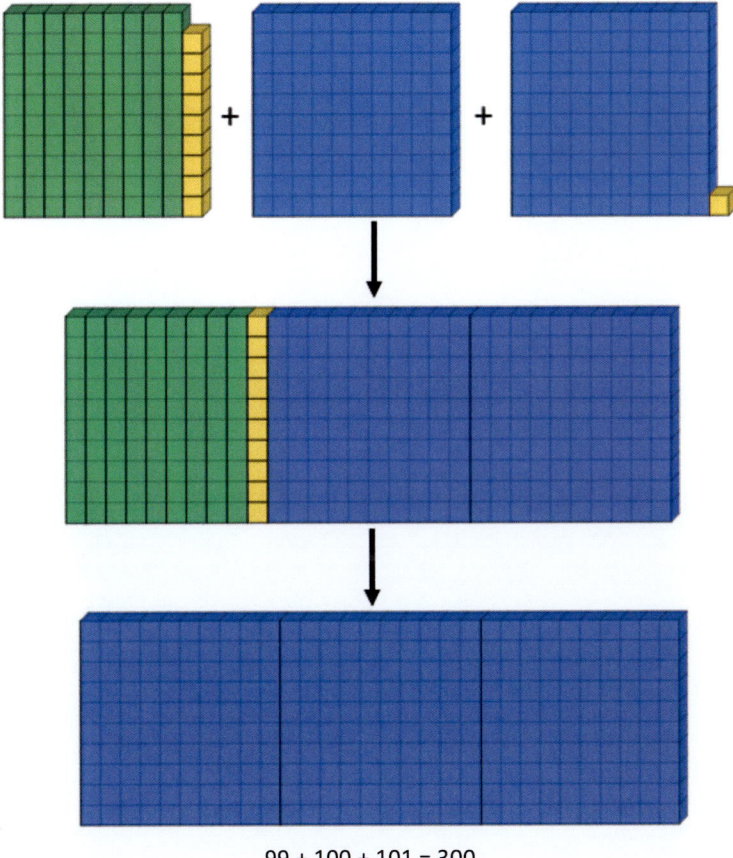

$$99 + 100 + 101 = 300$$

Algebraic Proof:

Let n be any positive whole number. Then, n+1 will be its consecutive number and n+1+1 will be the next consecutive number. Therefore, the sum of any 3 consecutive positive whole numbers will be:
n + (n + 1) + (n + 1 + 1) = 3n + 3

| The brackets are used to emphasize the consecutiveness of the numbers. |

Note that, 3n+3 = 3(n + 1), thus showing that the sum of 3 consecutive positive whole numbers is a multiple of 3.

Notes: Counters and Dienes used in the examples are from MathsBot.com and the Number Lines used in the examples are from Polypad.org.

Any other manipulative that shows unit value can be used instead of the counters, for example: maths link cubes.

Students may further investigate the relationship between three consecutive odd numbers and three consecutive even numbers to identify whether they yield a multiple of three or not.

First 20 multiples of 3: 3, 6, 9, 12, 15, 18, 21, 24, 27, 30, 33, 36, 39, 42, 45, 48, 51, 54, 57, 60.

> **AS010 Statement:** The difference between two numbers is always smaller than the minuend.
>
> **Answer:** SOMETIMES TRUE
>
> **Manipulatives:** Double-sided Counters

This statement can be shown as sometimes true through the use of *Double-sided Counters*.

Firstly, we consider two examples that demonstrate it as not true by considering the minuend to be a positive number and the subtrahend to be a negative number.

Example of the arrangement of double-sided counters shown in the diagram below is of the calculation 3 − −1 = 4. Here we employed the idea of zero pairs to support the thinking of taking away one number from another. As we cannot take away what we do not have, physically, we introduce zero pairs by adding them to the 3 yellow counters. This then allows for the physical act of taking away/ removing the negative 1 value as per the question.

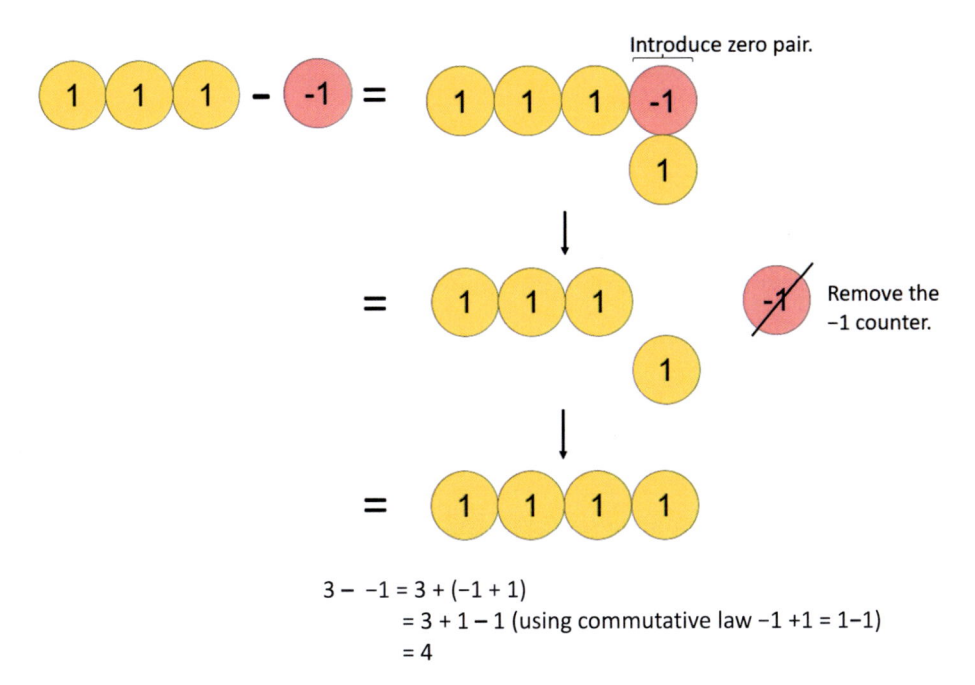

$$3 - -1 = 3 + (-1 + 1)$$
$$= 3 + 1 - 1 \text{ (using commutative law } -1 + 1 = 1 - 1\text{)}$$
$$= 4$$

The statement continues to show as not true when considering both the minuend and subtrahend to be negative numbers. For example, the arrangement of double-sided counters shown below is of the calculation −3 − −1 = −2. As the physical act of taking away is employed, we can simply 'take away' the −1 counter to reveal the result of −2. (−3 − −1 = −2)

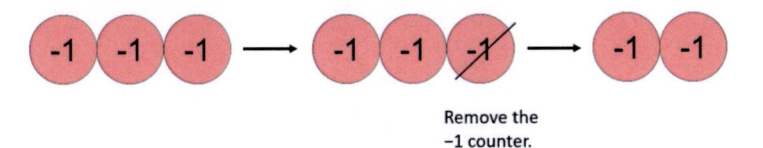

Now, we consider when the statement can be seen as true when considering both the minuend and subtrahend to be positive numbers.

Example of the arrangement of double-sided counters shown below is of the calculation where the minuend is larger than the subtrahend. As the physical act of taking away is employed, we can simply 'take away' the 1 counters to reveal the result.

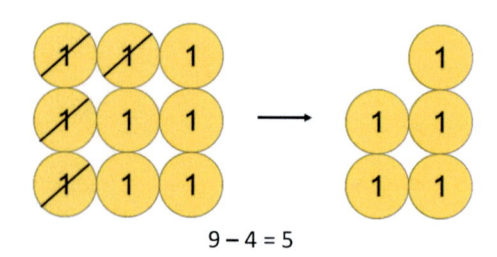

$$9 - 4 = 5$$

Example of the arrangement of double-sided counters shown below is of the calculation where the minuend is smaller than the subtrahend. As the physical act of taking away is employed, we need to introduce zero pairs to accommodate the removal of the counters with the value of the subtrahend. In the example, note that 5 zero pairs are introduced and not nine zero-pairs. This is because the subtrahend is of value 9. Therefore, as 9 is 5 more than 4, only 5 zero pairs need to be introduced. Example below shows 4 – 9 = –5. We 'take away' the nine 1 counters to reveal the result.

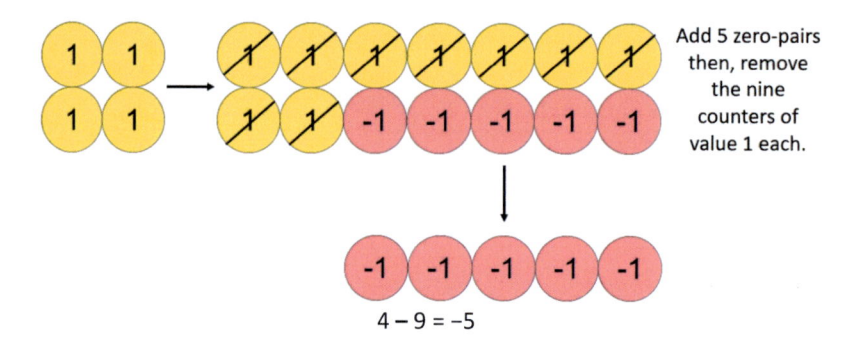

Add 5 zero-pairs then, remove the nine counters of value 1 each.

$$4 - 9 = -5$$

Hence, from all of the examples shown above, we can see that the statement is sometimes true.

> **Notes:** Double-sided Counters used in the diagrams are from MathsBot.com.
>
> Ensure that the students have clarity in the recording of the 'minus' sign. Is it an operation or a negative number? In the explanations the operation is shown in bold.
>
> Encourage the use of positive and negative numbers and the swapping of the value of the subtrahend and the minuend. What happens when zero is subtracted?
>
> Students may use number lines to investigate, however be aware of the common misconception of change in direction with subtractions and negative numbers.

AS011 Statement: When two numbers are added their order does not matter as the answer will be the same.

Answer: ALWAYS TRUE

Manipulatives: Cuisenaire Rods

This statement can be shown as always true through the use of *Cuisenaire rods* by physically joining the rods and by swapping the order of the two numbers being added. Diagrams that resemble the number bars as rectangles on square paper can be drawn instead, as well as writing number sentences to support the findings.

Examples of the use of number bars are shown below. The focus is on the length as the sum of the 2 chosen bars. By comparing the lengths shown below with one of the bars as added on and also as the one that is added to. With further investigation using different sized rods, the commutative property can be identified clearly. Notice there are no number values provided in the diagrams below. Students may assign numerical values to the bars by employing the idea of substitution. Can they generalise for any numerical value?

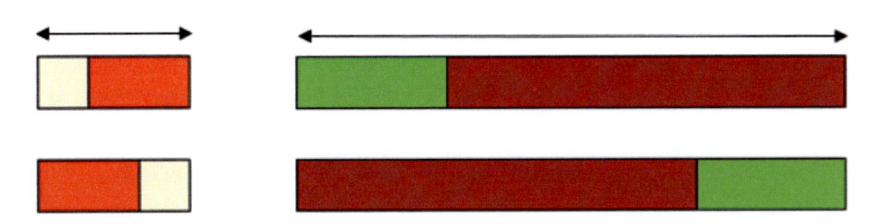

It can be seen from the total length of a pair of bars that the statement is always true.

Notes: Cuisenaire Rods used in these images are from MathsBot.com.

Understanding that the order of adding numbers together does not affect the total is important to support student's fluency. Sometimes, it can be more efficient to calculate the sum by placing the larger number first. For example: in 4 + 7 = __, calculate it as 7 + 4 = __ and can be thought of as 7 + 3 + 1 = __

Other manipulatives that show the total length does not change when swapping the numbers can be used, for example: number lines and maths link cubes.

Encourage students to investigate if the statement is true for numbers that are just odd numbers or just even numbers or a combination. See AS001 and AS003.

Also, students may further investigate if the same is true for subtraction. See AS014.

> **AS012 Statement:** Adding 2 to an even number results in an even number.
>
> **Answer:** ALWAYS TRUE
>
> **Manipulatives:** Number Frames

This statement can be shown as always true through the use of *Number Frames* by physically joining a number frame of value two to any other even number frame.

Diagrams that resemble the number frames (dots in square paper) can be drawn instead, as well as, writing number sentences to support the findings.

Examples of the use of number frames are shown below with the two frame being placed above the chosen even number frame as the act of adding. The evenness can be seen through equal pairing and the symmetry in the resulting sum.

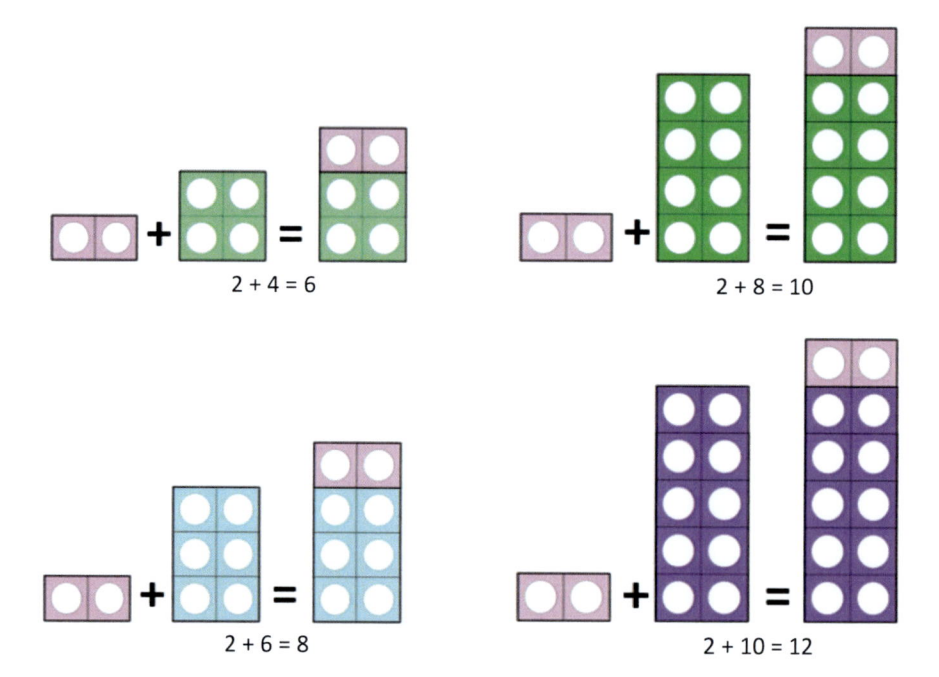

2 + 4 = 6

2 + 8 = 10

2 + 6 = 8

2 + 10 = 12

From the examples above, it can be concluded that the statement is always true.

Students may use other orientations of the number frames instead of those shown above. In that case, it may become visually tricky to see the symmetry.

Consider the creations shown in the diagrams on the right. These show that there is symmetry. However, visual understanding of halving one whole as well as the role that colour plays in identification of symmetry are required.

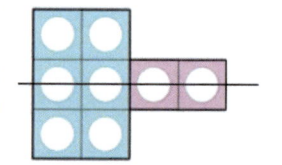

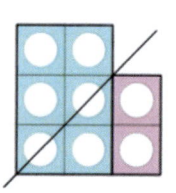

The diagram with the diagonal line of symmetry suggests symmetry of the shape but not through colour. On the other hand, the diagrams below show no symmetry thus, leading to believe that the number may not be an even number.

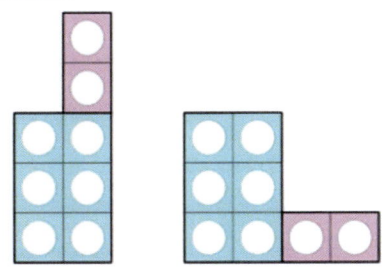

Therefore, when using number frames in different orientations, it provides an opportunity to discuss the symmetry of the sum and how rearrangement yields the evenness/ equal pairing.

> **Notes:** Number Frames used in these images are from Mathsbot.com.
>
> An opportunity to connect between subtraction and addition as an inverse relationship can be undertaken in tandem. As well as understanding commutativity with addition.
>
> Students may further predict what happens when even numbers are added to even numbers. This is discussed in AS003.
>
> All even numbers should be whole numbers. Even numbers have a ones (units) digit of either 0, 2, 4, 6 or 8.

> **AS013 Statement:** Adding 3 to an odd number results in an odd number.
>
> **Answer:** NEVER TRUE
>
> **Manipulatives:** Number Frames

This statement can be shown as never true through the use of *Number Frames* by physically joining a number frame of value three to any other odd number frame.

Diagrams that resemble the number frames (dots in square paper) can be drawn instead, as well as, writing number sentences to support the findings.

Examples of the use of number frames are shown below, where the three frame is placed such that it interlocks with the chosen odd number frame to reveal the sum. This shows the evenness of the sum resulting in the statement to be never true.

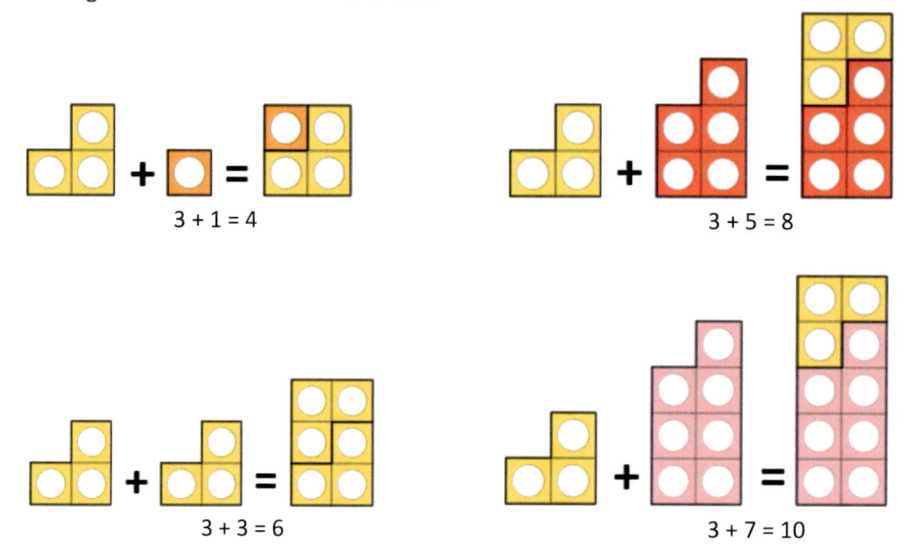

3 + 1 = 4

3 + 5 = 8

3 + 3 = 6

3 + 7 = 10

Students may use other orientations of the number frames instead of how it is shown above. In that case, it may become visually tricky to see the 'even number-ness'.

Consider the creations shown in the diagrams below. Even though (no pun intended), there may be an opportunity to see symmetry and how this connects with the evenness of a number, it emphasises the importance of discussing different orientations to further embed the idea of even number-ness.

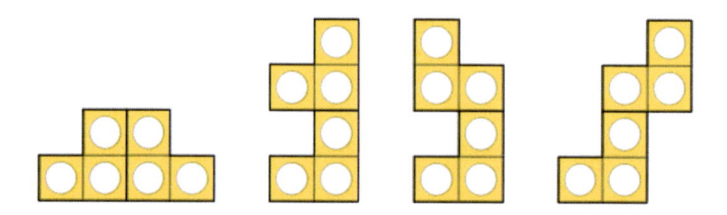

Notes: Number Frames used in these images are from MathsBot.com.

An opportunity to connect between subtraction and addition as an inverse relationship can be undertaken in tandem. As well as understanding commutativity with addition.

Students may further predict what happens when odd numbers are added to other odd numbers. This is discussed in AS001.

All odd numbers should be whole numbers. Odd numbers have a ones (units) digit of either 1, 3, 5, 7 or 9.

AS014 Statement: Swapping the order of two different numbers when finding their difference will result in the same answer.

Answer: NEVER TRUE

Manipulatives: Dienes, Number Lines

Using Dienes

This statement can be shown as never true through the use of *Dienes* by removing the required number in the calculation – the subtrahend from the minuend (Note: minuend – subtrahend = difference). Followed by seeing what happens when the subtrahend and minuend are swapped with each other – such that the subtrahend becomes the minuend, and the minuend becomes the subtrahend. Writing number sentences alongside the practical resources will support the findings.

First let's consider using a single digit. In the first example 7 – 3, it is possible to remove 3 unit cubes. However, when swapped to 3 – 7, students will notice that they can remove 3 but not a further 4 (shown here with the lines with no cubes underneath them). Practical example such as "If you have 3 apples, can you remove 7 from them?", can support making sense of the idea.

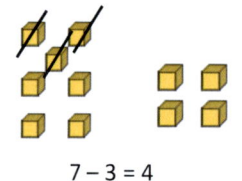

$$7 - 3 = 4 \qquad 3 - 7 = -4$$

A further example shown with a two-digit number is shown below.

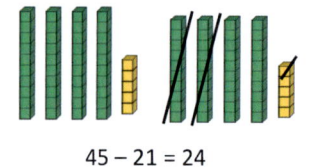

$$45 - 21 = 24 \qquad 21 - 45 = -24$$

(Shown by the lines without cubes underneath them)

Both the examples suggest that the statement is never true. Encourage students to explore by using a wide range of numbers.

Using Number Lines

Using the numbers 7 and 3 as in the example above, the first diagram below shows the start of the arrow on the 7 and jumping back 3 spaces pointing to 4 as the answer (7 – 3 = 4)

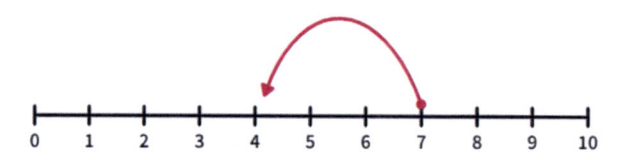

Whereas, starting at 3 and jumping back 7 spaces points to ‾4 as the answer. These are different places on the number line. (3 – 7 = –4)

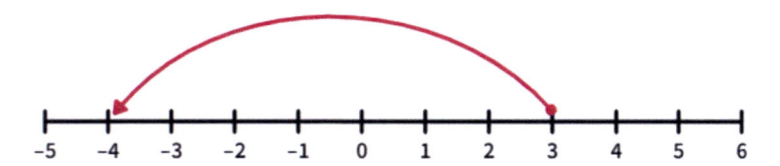

Encourage the students to investigate if the statement is true for numbers that are odd and even, including a combination of some single digit and two or three-digit numbers.

Again, it can be seen that the statement is never true.

Note that the condition placed with this statement is that the numbers cannot be the same. Would the statement remain to be never true?

> **Notes:** Dienes used in the diagrams are from MathsBot.com and the Number Lines used in the diagrams are from Polypad.org.
>
> Another manipulative that could be used is double-sided counters. See AS010 for a similar example using double-sided counters.
>
> The commutative law is an important rule in mathematics and says that numbers can be swapped around when you add them or multiply them and the same answer will still result, but this does not apply when subtracting or dividing.

> **AS015 Statement:** The sum of two 2-digit numbers will be a 3-digit number.
>
> **Answer:** SOMETIMES TRUE
>
> **Manipulatives:** Dienes

This statement can be shown as sometimes true through the use of *Dienes* by physically joining two 2-digit whole number representations. Here, the familiarity of a 100 frame (10 by 10 frame) should be encouraged to be used as a comparison with the sum to identify whether the sum is less than or equal to or greater than 100 (the smallest 3-digit number).

Diagrams that resemble the Dienes in 2D can be drawn on squared paper instead, as well as, writing number sentences to support the findings.

Encourage the students to investigate several different number combinations to identify when the statement is true and when it is not true.

Examples of the arrangements of the Dienes are shown in the diagrams below, where the statement is **not true**. Notice the sentences written underneath the diagram to show the understanding and connect the visual with the abstract.

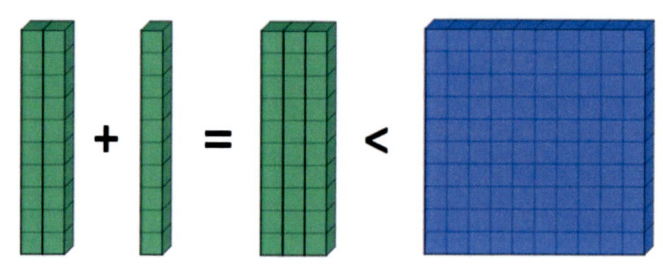

20 + 10 = 30
30 < 100 Therefore, in this case, the statement is not true.

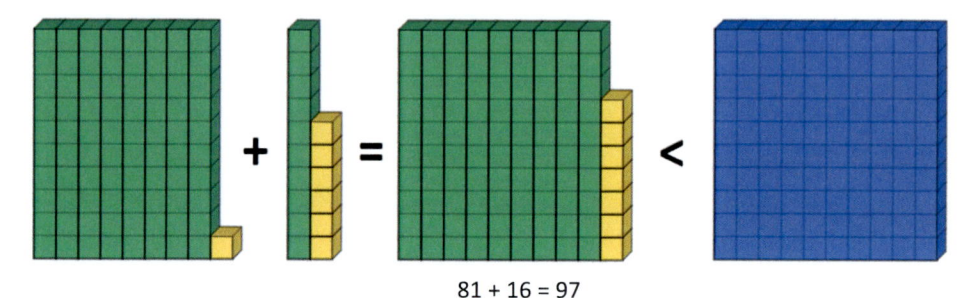

81 + 16 = 97
97 < 100 Therefore, in this case, the statement is not true.

Whereas, in the diagram on the following page, the statement is **true**. Notice how the 10 Tens Dienes have been regrouped for the 100 square. The number sentence supports the findings, including the statement with the inequality.

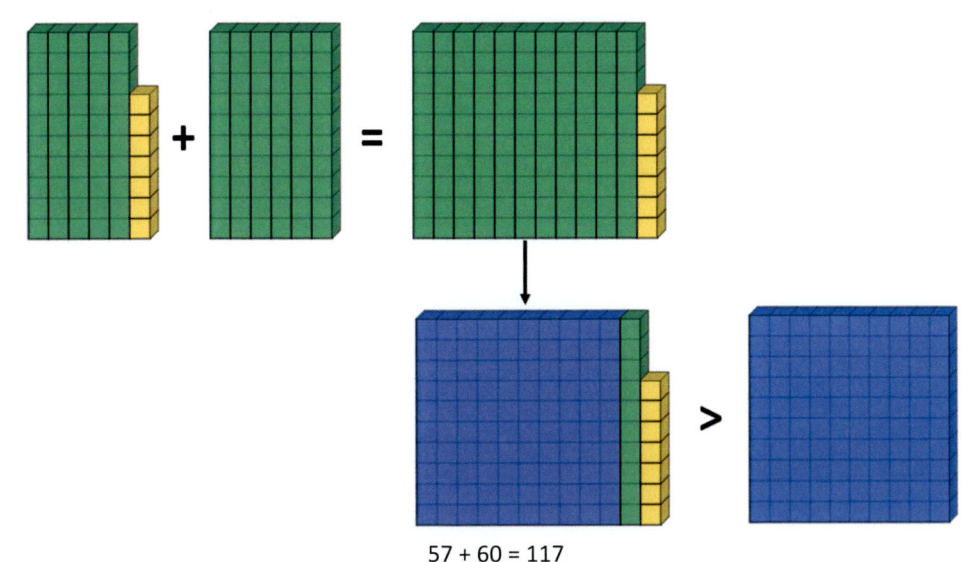

57 + 60 = 117
117 > 100 Therefore, in this case, the statement is true.

From the examples above, the statement is sometimes true.

Notes: Dienes used in the images are from MathsBot.com.

Cuisenaire rods or maths link cubes can be used as an alternative manipulative.
Students may use different arrangements, in which case, a focus point can be on the visual representation of the sum compared to that of 100 to support their deductions.

Be aware of situations where children have worked with lots of different values but, none whose sum is 100 or more and vice versa. Ask questions like - How do we know that we have considered all possibilities before making a generalisation?

2-digit numbers are numbers from 10 up to and including 99.
3-digit numbers are numbers from 100 up to and including 999.

AS016 Statement: 50 can be made by adding three odd numbers.

Answer: NEVER TRUE

Manipulatives: Rekenrek

This statement can be shown as never true through the use of a *Rekenrek* by moving the odd number of beads to the left and where necessary, to rearrange them to visually compare whether the sum of the 3 odd numbers is less than, equal to or greater than 50.

The diagrams below show how the rekenrek visually displays the sum of three 9s to be an answer less than 50, the sum of 21, 9 and 15 to be an answer less than 50, the sum of 23, 11 and 17 to be an answer greater than 50. Hence, showing the statement is not true in these instances.

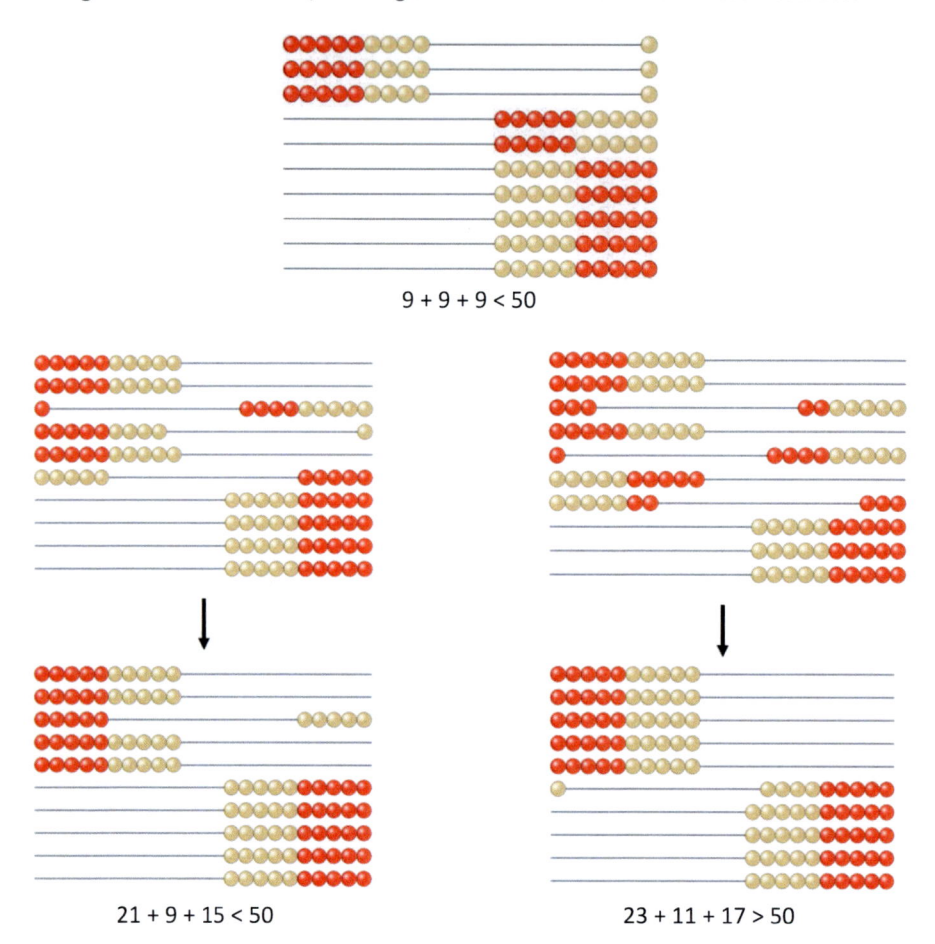

9 + 9 + 9 < 50

21 + 9 + 15 < 50

23 + 11 + 17 > 50

Diagrams that resemble the rekenrek like dots or circles on squared paper can be drawn instead. In this case, ensure that the diagrams created are comparable in groups or rows of 10.

Alternatively, this statement can be shown as never true through the use of writing the sum of two odd numbers and an even number to show that an even number is required for the sum to be 50.

Thus employing the idea of proof by contradiction.

As 50 is an even number, students may predict that no combination of odd numbers will result in their sum to equal 50.

Using a systematic approach in writing the number sentences reveals the pattern that when two odd numbers are added, their sum is always an even number. Through the cumulative addition, students can use the knowledge of adding even numbers yields an even number. (See AS001, AS003 and AS005).

An example of the systematic presentation of number sentences is shown below.

Odd + Odd + Even = 50
1 + 1 + 48 = 50
1 + 3 + 46 = 50
1 + 5 + 44 = 50
1 + 7 + 42 = 50
1 + 9 + 40 = 50
3 + 3 + 44 = 50
3 + 5 + 42 = 50
3 + 7 + 40 = 50
3 + 9 + 38 = 50
5 + 5 + 40 = 50
5 + 7 + 38 = 50
5 + 9 + 36 = 50
7 + 7 + 36 = 50
7 + 9 + 34 = 50
9 + 9 + 32 = 50

> **Notes:** Rekenrek used in the images are from MathsBot.com.
>
> A metre stick or measuring tape can also be used as a number line to visualise the cumulative addition and identify it as whether it is less than, equal to or greater than the 50 marked on them.
>
> Number frames, maths link cubes or Dienes can be used as alternative manipulatives. If these are used then, discuss the impact of orientation as the students explore the statement.
>
> This investigation can be linked to 'The sum of three odd numbers is always odd'. Thus, concluding that the answer in this case can never be an even number. See AS007.
>
> All odd and even numbers should be whole numbers. Odd numbers have a ones (units) digit of either 1, 3, 5, 7 or 9. Even numbers have a ones (units) digit of 0, 2, 4, 6 or 8.

> **AS017 Statement:** Adding 9 to a number is the same as adding 10 and subtracting 1.
>
> **Answer:** ALWAYS TRUE
>
> **Manipulatives:** Dienes

This statement can be shown as always true through the use of *Dienes* by manipulating with them through the movement of a ten and the appropriate number of ones depending on the numbers involved. Note that the idea of subitising may be employed to support the discussions as comparisons between arrangements of unit cubes are made.

Example using a single digit, and the use of Dienes are shown below alongside respective number sentences to support the findings.

7 + 9 = 16

7 + 10 − 1 = 16

A further example shown with a two-digit number is shown here:

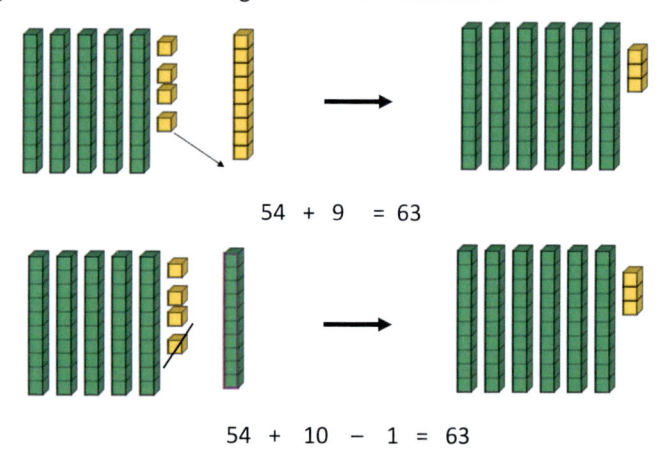

54 + 9 = 63

54 + 10 − 1 = 63

From the examples, it can be seen that the statement is always true. Encourage the students to investigate if the statement is true for numbers that are odd and even, some single digit and some two or three-digit numbers.

> **Notes:** Dienes used in these images are from MathsBot.com.
>
> Some other manipulatives that could be used are number lines and maths link cubes.
> Also, students may further investigate if the same is true for addition of 99 as adding 100 and subtracting one. Similarly with adding 999 by considering adding 1000 and then subtracting one.

> **AS018 Statement:** Adding 10 to a number results in a number that is a multiple of 10.
>
> **Answer:** SOMETIMES TRUE
>
> **Manipulatives:** Cuisenaire Rods, Place Value Cards

Using Cuisenaire Rods

This statement can be shown as sometimes true through the use of *Cuisenaire rods* by visually comparing the length of the rod that represents the selected number to the length of the multiples of 10.

From the examples shown below, it can be seen that the cases where the statement is true are for the numbers 0 and 10.

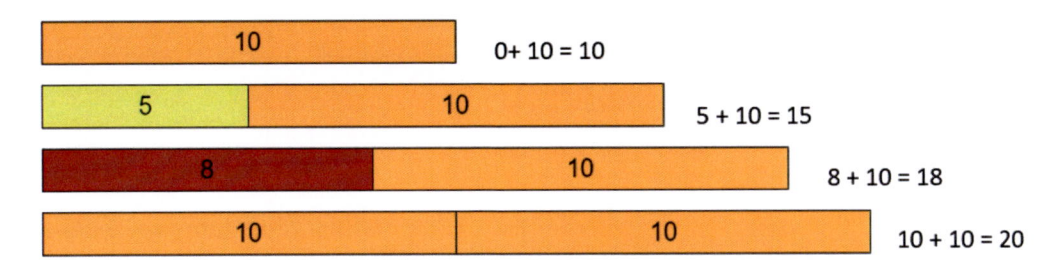

10	0 + 10 = 10
5	10	5 + 10 = 15
8	10	8 + 10 = 18
10	10	10 + 10 = 20

Encourage students to explore a variety of different numbers to support their predictions. What do they notice?

Using Place Value Cards

Place Value Cards can provide another visual comparison through the movement of the cards according to the place value. Here the assumption is the understanding of the change in the tens place value when adding a 10 to any given number.

In the examples below, the place value cards are utilised like displaying the formal column addition method without displaying the addition operation but performing the operation.

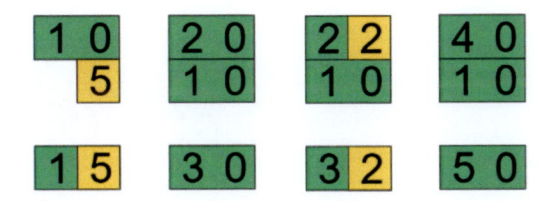

1 0	2 0	2 2	4 0
5	1 0	1 0	1 0
1 5	3 0	3 2	5 0

Encourage students to discuss what is the same and what is different. What do they notice? Are the outcomes the same when compared to the Cuisenaire rods?

From the examples shown the statement is sometimes true.

Notes: Cuisenaire Rods and Place Value cards used in the diagrams are from MathsBot.com.

Notice that the examples provided here are for positive integers. Understanding that rearranging the order of the numbers (commutativity) will not affect the answer can also be drawn out through this investigation. Also, students may further investigate what happens when they start with a negative number.

Other manipulatives that show the length as the constant when rearranging the numbers can be used, for example: number lines and maths link cubes.

Is there a connection with odd and even numbers?

> **AS019 Statement:** The sum of two numbers is the same as the product of those numbers.
>
> **Answer:** SOMETIMES TRUE
>
> **Manipulatives:** Cuisenaire Rods

This statement can be shown as sometimes true through the use of *Cuisenaire rods* by physically joining them according to the interpretation of the operation. Understanding of area is beneficial, however discussion about the physical visual comparisons are essential.

Diagrams that resemble the rods like bars on squared paper can be drawn instead, as well as writing number sentences to support the findings.

Examples of the arrangements of the Cuisenaire rods are shown below, where the statement is not true. Notice the sentences written underneath the diagram to show the understanding and connect the visual with the abstract. Also, notice that a different colour is used to represent the area as the product of the two numbers in question. An alternative way of showing the area is through identifying the side lengths.

Example 1:

Adding two numbers:
1 and 1

$1 + 1 = 2$

Area model to show the multiplication of the same two numbers:

$1 \times 1 = 1$

Alternative representation:

$1 \times 1 = 1$

As, $1 + 1 = 2 \neq 1 \times 1$ Therefore, in this case, the statement is not true.

Example 2:

Adding two numbers:
2 and 2

$2 + 2 = 4$

Area model to show the multiplication of the same two numbers:

$2 \times 2 = 4$

Alternative representation:

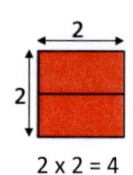

2 × 2 = 4

As, 2 + 2 = 4 = 2 × 2   Therefore, in this case, the statement is true.

Example 3:

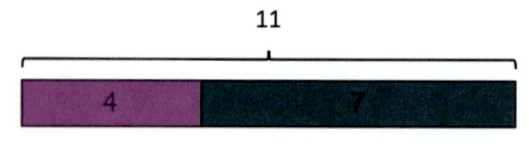

Adding two numbers:
4 and 7

4 + 7 = 11

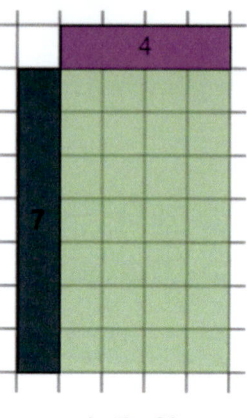

Area model to show the multiplication of the same two numbers:

4 × 7 = 28

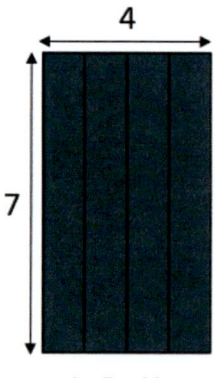

Alternative representation:

4 × 7 = 28

As, 4 + 7 = 11 ≠ 4 × 7   Therefore, in this case, the statement is not true.

From the examples above, it can be deduced that the statement in sometimes true. Students may explore if the statement is always true for numbers that are the same, as in example 2 above.

Students may use different orientations, and these can be used to discuss the commutative law with addition as well as multiplication.

Notes: Cuisenaire Rods used in the diagrams are from MathsBot.com and Number Lines used in the diagrams below are from Polypad.org.

Number lines and unit cubes such as maths link cubes can be used as alternative manipulatives. If using number lines, be clear with the representation of multiplication 'as groups of ' relating to the number of jumps on the number line. For example: 1 + 1 represented as 2 whereas, 1 x 1 is represented as 1 group of 1.

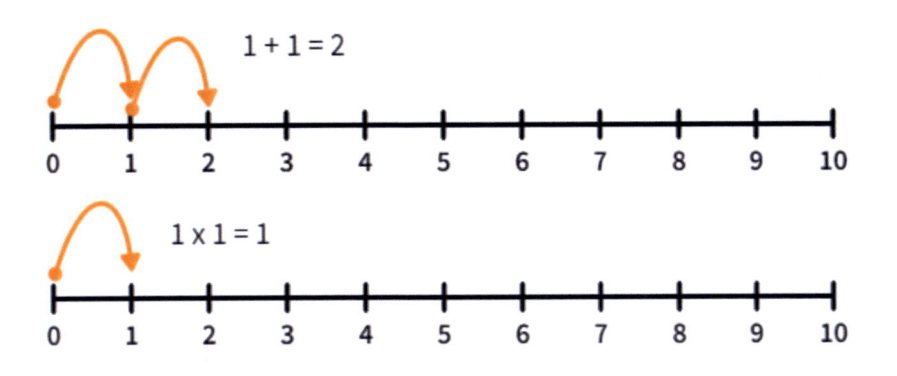

1 + 1 = 2

1 x 1 = 1

> **AS020 Statement:** The difference between a 2-digit number and its' reverse is a multiple of 9.
>
> **Answer:** ALWAYS TRUE
>
> **Manipulatives:** Rekenrek and Number Line

This statement can be shown as always true through the use of a *Rekenrek* by starting with moving the beads to the left to show the minuend. Then, showing the action of taking away as moving the beads in the quantity of the subtrahend from the minuend to the right. This will reveal the difference.

The diagrams below show how the rekenrek visually displays the value of the difference with the minuend shown as the starting point. As it is possible to move the beads for the subtrahend one by one or in groups of ten, visualisation can be supported by using a number line to identify that the answer is always a multiple of 9.

Example 1:
Let's consider the number 11. In this diagram, the value of the difference is shown by the representation followed by the arrow. In this case, it is zero as there are no beads on the left in the rekenrek display. (11 – 11 = 0)

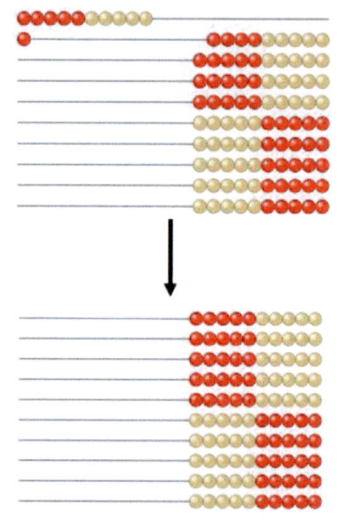

The above calculation can be shown using a number line as seen here:

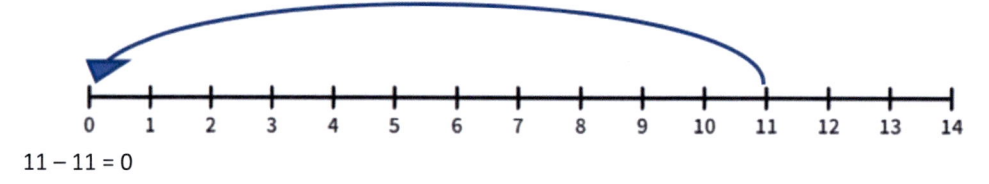

11 – 11 = 0

This is valid as 0 x 9 = 0. This may need further discussion if students do not connect with the value of zero in the multiplication table.

Example 2:
Now let's consider the number 21. Again, the value of the difference is shown by the representation followed by the arrow. In this case, it is 9 as seen by the number of beads on the left in the rekenrek display below. (21 – 12 = 9)

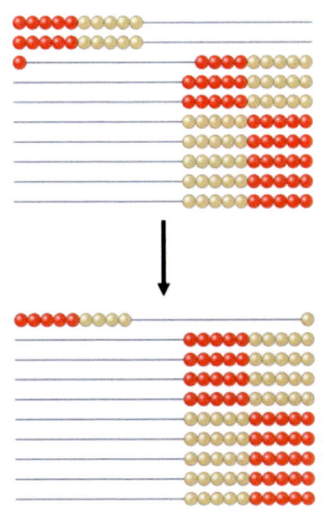

The above calculation can be shown using a number line as seen here:

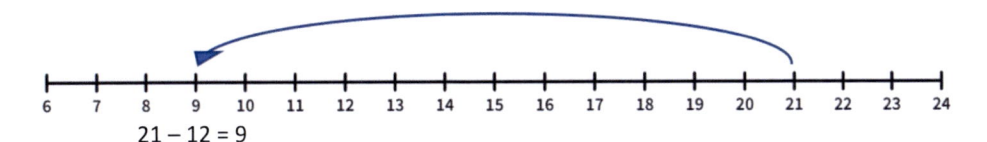

21 – 12 = 9

This is valid as 1 x 9 = 9

Example 3:
Finally let's consider the number 62. In the diagram here, the value of the difference can be seen as 36 by the number of beads on the left in the rekenrek display.
(62 – 26 = 36)

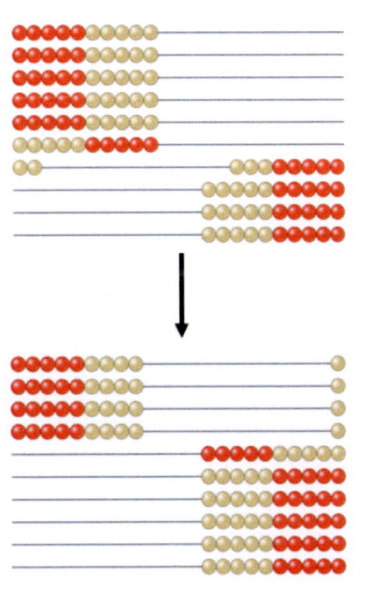

Note that students may move the beads from right to left using a different strategy. The focus should remain on the final value of the difference, although a visual display of multiples of 9 would be more convincing to the eye.

The above calculation can be shown using a number line, albeit the miniscule digits:

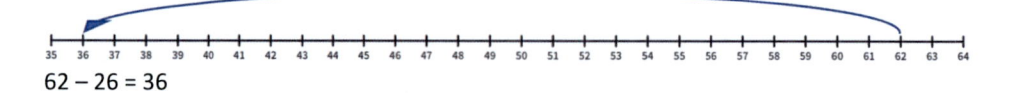

62 – 26 = 36

This is valid as 4 x 9 = 36

From the above examples, the statement is always true. Encourage the students to connect their knowledge of the divisibility rule for the multiplication table of 9 to support their explanations as to why the statement is always true.

> **Notes:** Rekenreks used in diagrams are from MathsBot.com and the Number Lines used in diagrams are from Polypad.org.
>
> A metre stick or measuring tape can also be used as a number line to support 2-digit numbers greater than 30.
>
> The subtraction method used with the number lines is that of taking away rather than finding the difference by starting with the subtrahend and stating how much more is needed to reach the minuend.
>
> Students can investigate further by finding the difference between 3-digit numbers and more. For example: 321 – 123 = 198. Applying the divisibility rule reveals that the digits 198 can be added to reveal 1 + 9 + 8 = 18, which can further be added to reveal 1 + 8 = 9, thus showing that the difference is a multiple of 9.
>
> An in-depth proof can be seen in the famous 1089 problem described by David Acheson.

AS021 Statement: The sum of four even numbers is divisible by 4.

Answer: SOMETIMES TRUE

Manipulatives: Number Frames

This statement can be shown as sometimes true through the use of *Number Frames* by physically joining four even number frames together. Understanding of groups of 4 is essential to support the visual identification of the sum being divisible by 4.

Diagrams that resemble the number frames like dots or circles on squared paper can be drawn instead, as well as, writing number sentences to support the findings.

An example of the arrangement of four different number frames is as shown in the diagrams below. Notice the exchange of the pieces to highlight the four-ness as well as the exchange with decomposing one of the pieces to show groups of 4.

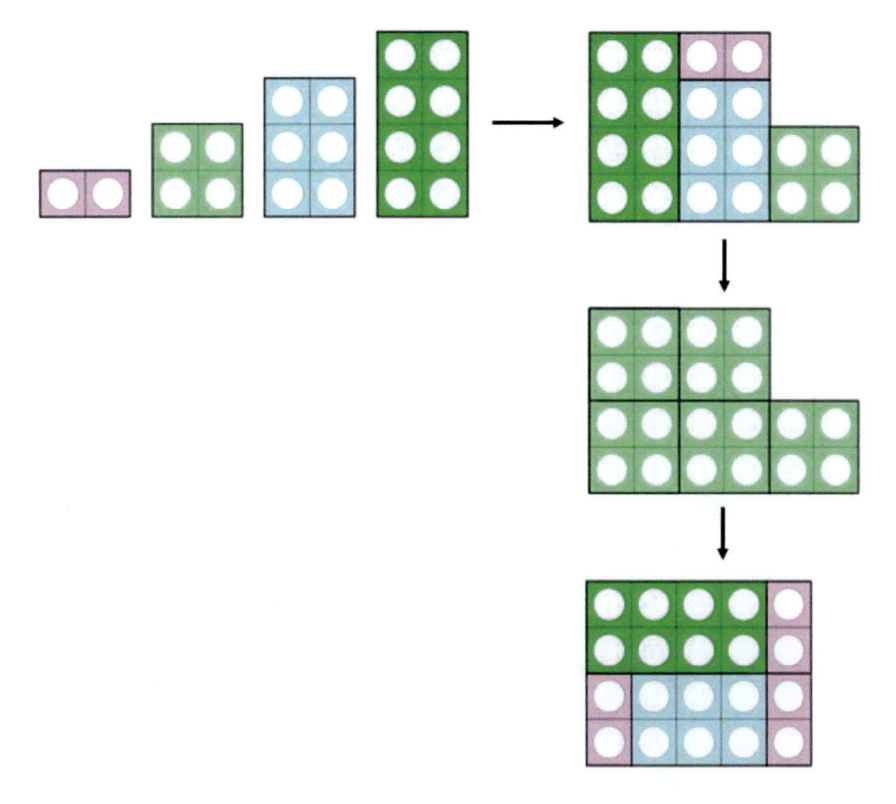

$$2 + 4 + 6 + 8 = 20 \quad \text{and} \quad 20 \div 4 = 5$$

For the example above, the physical manipulation with the number frames and the calculations appear to reveal the statement is always true. Is it true for all cases? Encourage students to explore through a range of different even numbers as well as same even numbers. What relationships do they notice?

Students may use different orientations and identify different ways of decomposing their numbers to 'make' a shape that reveals its four-ness readily.

As students explore with a combination of even numbers, there are exceptions for which this statement is not true. Consider the numbers 0 and 2 in this sum: 0 + 0 + 0 + 2 = 2. As it is not possible to show a number frame of value 0, we consider representing the sum: 2 and comparing it to the representation of 4.

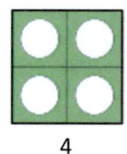

2   4

The visual representation shows us that 2 is not divisible by 4, indeed it is smaller than 4. Similarly, 0 + 0 + 0 + 6 = 6. Even though 6 is greater than 4, it is also not divisible by 4.

Encourage students to discuss this using their knowledge about multiples or the divisibility rule of 4. How can this finding be articulated? Are there other cases, where the statement is never true?

From both examples above, the statement is sometimes true.

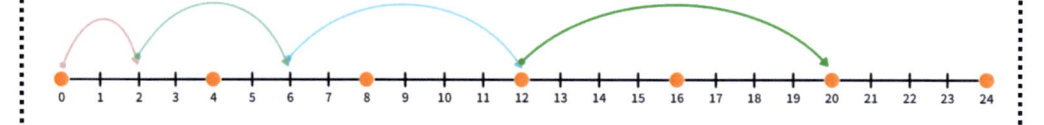

> **Notes:** Number Frames used in the diagrams are from MathsBot.com and the Number Line used in the diagram below is from Polypad.org.
>
> Number lines, Cuisenaire rods and unit cubes such as maths link cubes can be used as alternative manipulatives. If using number lines then, mark the number line with the multiples of 4 prior to showing the addition of the four even numbers as in the example here: 2 + 4 + 6 + 8 = 20

> **AS022 Statement:** The square of an even number is divisible by 4.
>
> **Answer:** ALWAYS TRUE
>
> **Manipulatives:** Number Frames, Number Lines, Cuisenaire Rods

Using Number Frames

This statement can be shown as always true through the use of *Number Frames* by physically creating a square structure using even number frames and rearranging the frames such that it is displayed in groups of 4. Thus supporting the connection between division and multiplication.

Examples of the arrangements of number frames are shown in the diagrams below. Notice that the products are displayed in a vertical grouping of 4.

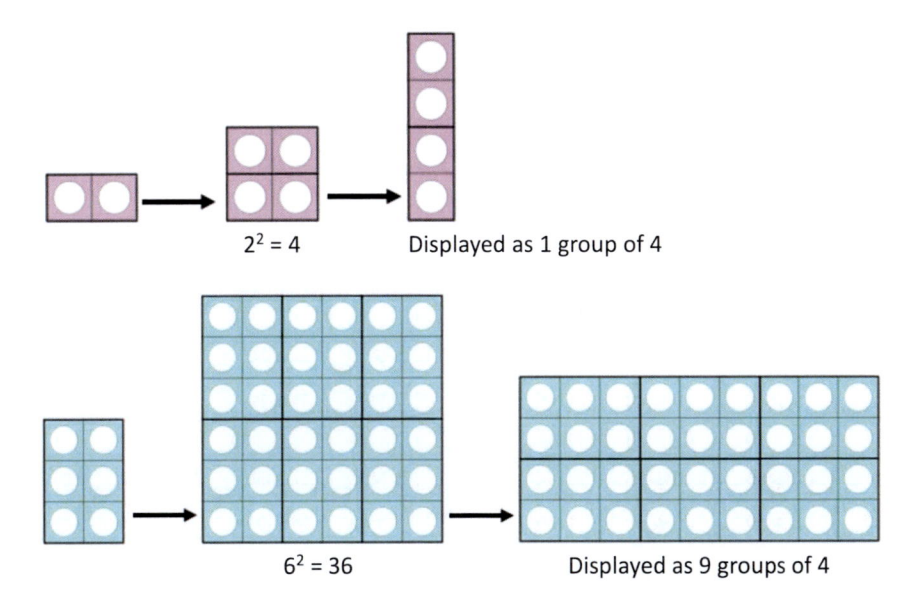

$2^2 = 4$ Displayed as 1 group of 4

$6^2 = 36$ Displayed as 9 groups of 4

From the above examples, the statement is always true. Encourage the students to explore a range of values. Students may use the horizontal groups of 4 instead of the vertical groups of 4 shown in the examples. Again, an opportunity here to discuss the commutative property of multiplication. Ensure that their choice of display is vivid in the grouping of 4 to support their findings.

Using Number Lines

Alternatively, a *Number line* can be used where the square number is identified as 'the number at the end of the jumps'. Here, the understanding is that the number of jumps is equivalent to the even number under investigation. This representation connects repeated addition to multiplication.

The diagrams on the following page show the jumps from squaring the even number 2 and 6. In both number lines notice the mark making of the number line with coloured dots to identify multiples of 4. This is done prior to using the number line. It is visible on the number lines that the square of an even number is a multiple of 4. As it is identified as a multiple of 4, through multiplicative reasoning, it can be inferred that it is divisible by 4.

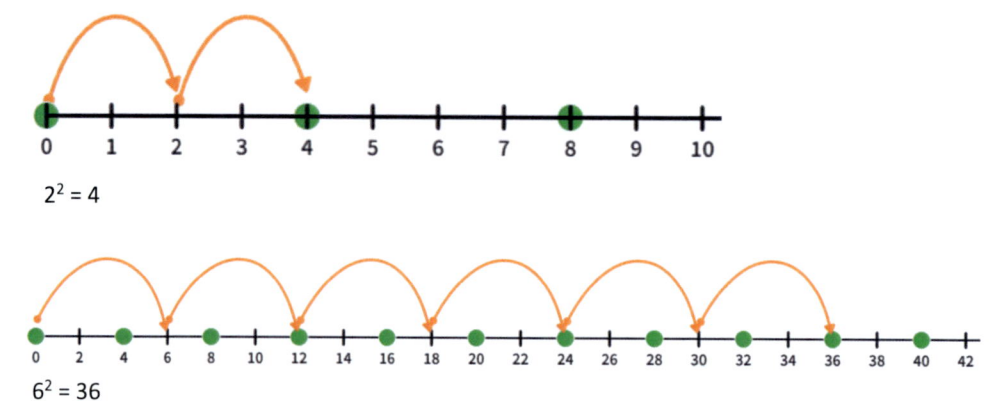

$2^2 = 4$

$6^2 = 36$

As expected, these reveal the same results as before, appearing as though the statement is true.

Using Cuisenaire Rods

Cuisenaire rods can be arranged in the form shown in the diagrams below, such that the area bound by the even numbers uses rods of value 4 or shows a grouping of 4.

Notice that the 10 by 10 square doesn't fit the rods as neatly as the 8 by 8 square however, it does accommodate for groups of 4. It is the whole group of 4 that is identified as the connection with divisible by 4. Therefore, it is acceptable to say that 100 is divisible by 4.

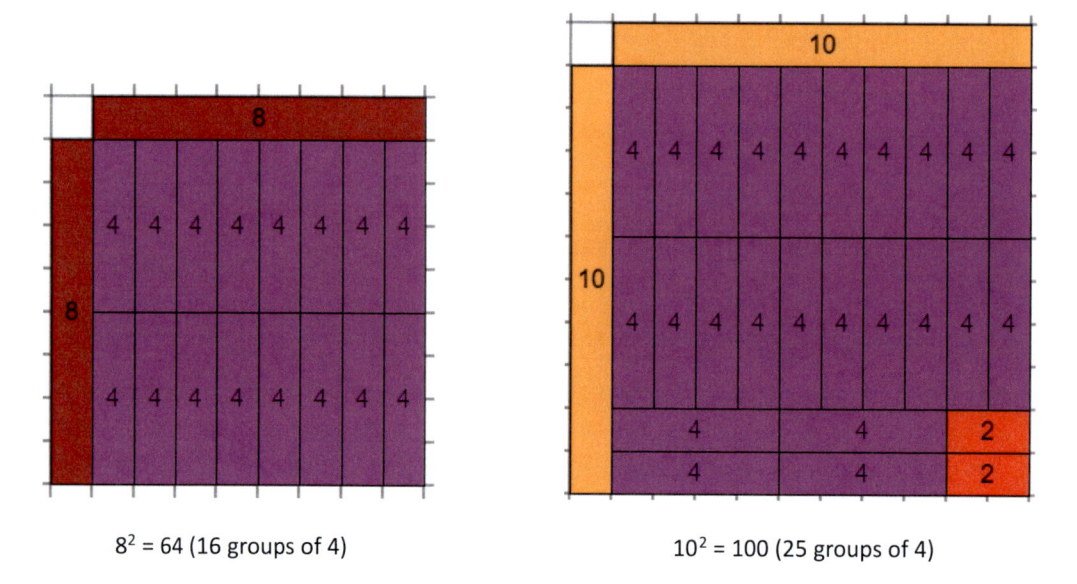

$8^2 = 64$ (16 groups of 4)

$10^2 = 100$ (25 groups of 4)

From all the examples above, the statement is always true.

Notes: Number Frames and Cuisenaire Rods used in the examples are from MathsBot.com and the Number Lines are from Polypad.org.

Any other manipulative that shows unit value can be used instead of the number frames, for example: maths link cubes or Dienes.

Students may further explore a similar relationship with the square of an odd number.

First 20 square numbers:
1, 4, 9, 16, 25, 36, 49, 64, 81, 100, 121, 144, 169, 196, 225, 256, 289, 324, 361, 400.

First 20 multiples of 4:
4, 8, 12, 16, 20, 24, 28, 32, 36, 40, 44, 48, 52, 56, 60, 64. 68. 72, 76, 80.

First 20 even number:
2, 4, 6, 8, 10, 12, 14, 16, 18, 20, 22, 24, 26, 28, 30, 32, 34, 36, 38, 40.

AS023 Statement: The sum of two triangular numbers is a square number.

Answer: SOMETIMES TRUE

Manipulatives: Counters

This statement can be shown as sometimes true through the use of *Counters* by physically arranging them so that they form a square shape. Students should be familiar with the physical representation of a square number as the shape of a square.

Diagrams that resemble the counters like dots or circles on square paper can be drawn instead, as well as, writing number sentences to support the findings.

Examples of the use of counters are shown in the diagrams below. Notice that the first diagram shows the sum of two triangular numbers as true. These numbers are consecutive triangular numbers. Will the statement be always true for consecutive triangular numbers?

1 + 3 = 4

Whereas, the diagram below shows that the sum of two triangular numbers is not true. Therefore, supporting the argument that if the triangular numbers are not consecutive then, their sum will not be a square number.

1 + 6 = 7

However, let's consider the numbers 1 and 15:

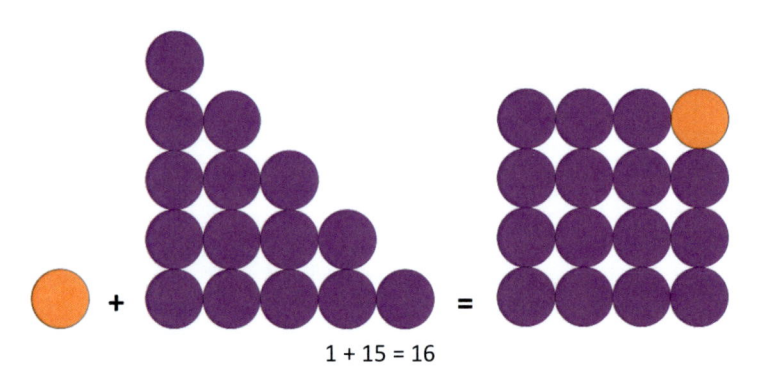

1 + 15 = 16

These are not consecutive triangular numbers, however they do make a square number, 16. Therefore, the statement is sometimes true. Hence, encourage the children to investigate more than just three different examples to support the identification of the relationship between consecutive triangular numbers and those that are not consecutive.

Students may use other orientations of the triangular numbers instead of those shown in the diagrams above. It is important to discuss the orientations and understand whether they make a difference to the structure or not. Thus, supporting the connection between the shape and its respective number.

> **Notes:** Counters used in the examples are from MathsBot.com.
>
> Students may further predict what happens when more triangular numbers are added. Thus, providing further investigative opportunities. See AS024.
>
> First 20 Triangular numbers:
> 1, 3, 6, 10, 15, 21, 28, 36, 45, 55, 66, 78, 91, 105, 120, 136, 153, 171, 190, 210.
>
> First 20 Square numbers:
> 1, 4, 9, 16, 25, 36, 49, 64, 81, 100, 121, 144, 169, 196, 225, 256, 289, 324, 361, 400.

AS024 Statement: Square numbers can be made by adding three triangular numbers.

Answer: SOMETIMES TRUE

Manipulatives: Counters

This statement can be shown as sometimes true through the use of *Counters* by physically arranging them so that they form a square shape structure. The use of the counters supports the visualisation of the square structure even if the square number is unknown by the children.

Diagrams that resemble the counters like dots or circles on square paper can be drawn instead, as well as, writing number sentences to support the findings.

Examples of the use of counters are shown in the diagrams below. Notice that the first diagram shows that a square number cannot be created from three triangular numbers. Hence, making the statement not true for this situation. These three numbers happen to be consecutive triangular numbers. Can students find a set of three consecutive triangular numbers that validates the statement? For example: the consecutive triangular numbers 15, 21 and 28, do indeed obtain a square number: 64.

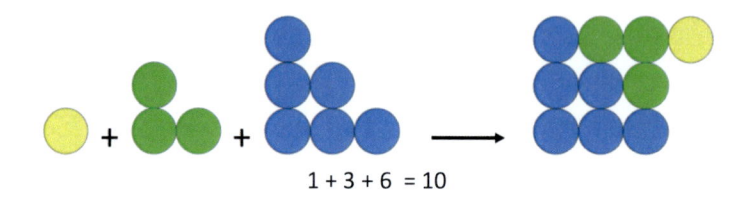

$$1 + 3 + 6 = 10$$

Whereas, the image below shows that a square number can be created from three triangular numbers. Hence, making the statement true for this situation. Notice that these three numbers are not consecutive.

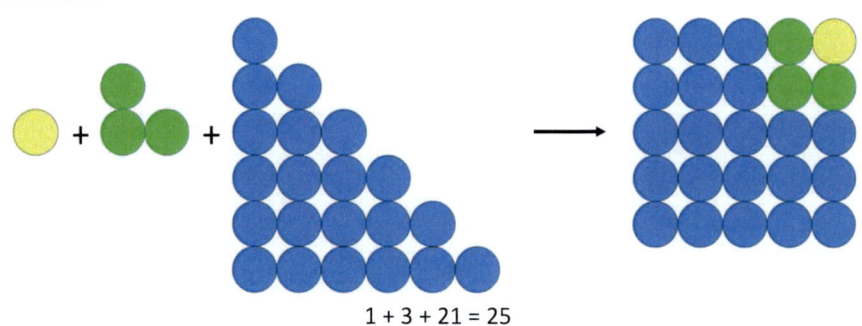

$$1 + 3 + 21 = 25$$

The above examples do concur that the statement is sometimes true. Are there any conditions for which the statement is true?

Students may use other orientations of the triangular numbers instead of those shown in the diagrams above. It is important to discuss the orientations and understand whether they make a difference to the structure or not. Thus, supporting the connection between the shape and its respective number.

Notes: Counters used in the diagrams are from MathsBot.com.

See AS023 for investigating the sum of two triangular numbers.

Students may further investigate the relationship between an odd number and even number of consecutive triangular numbers in terms of whether they yield a square number or not.

First 20 Triangular numbers:
1, 3, 6, 10, 15, 21, 28, 36, 45, 55, 66, 78, 91, 105, 120, 136, 153, 171, 190, 210.

First 20 Square numbers:
1, 4, 9, 16, 25, 36, 49, 64, 81, 100, 121, 144, 169, 196, 225, 256, 289, 324, 361, 400.

AS025 Statement: The sum of adding 4 to a negative number is a positive number.

Answer: SOMETIMES TRUE

Manipulatives: Number Lines, Double-sided Counters

Using Number Lines

This statement can be shown as sometimes true through the use of *Number Lines* by keeping a constant jump of 4 as a movement to the right of the number line (as a motion of adding) from any given number.

In the examples below, the result (the sum) shows some values as positive numbers and others as negative numbers.

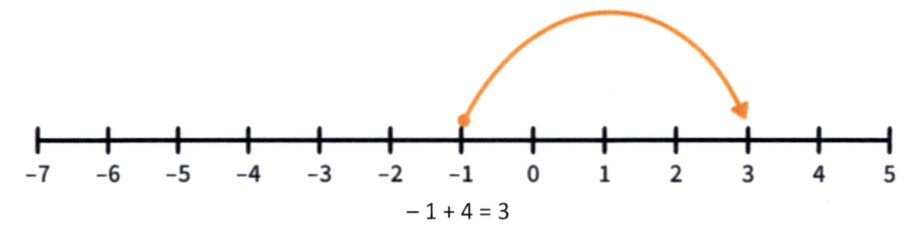

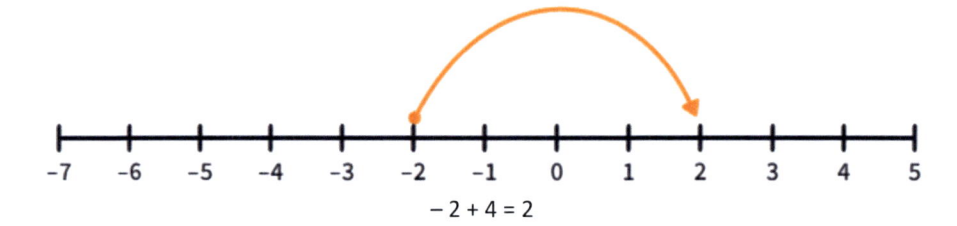

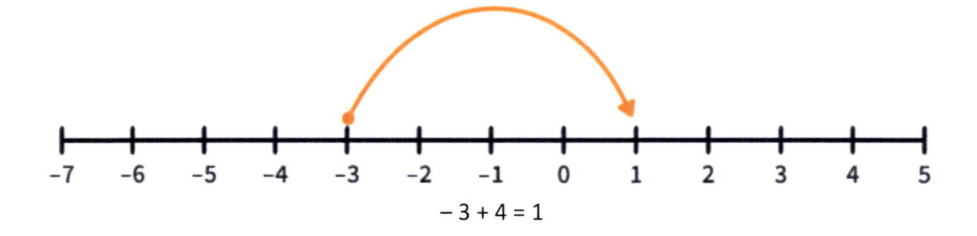

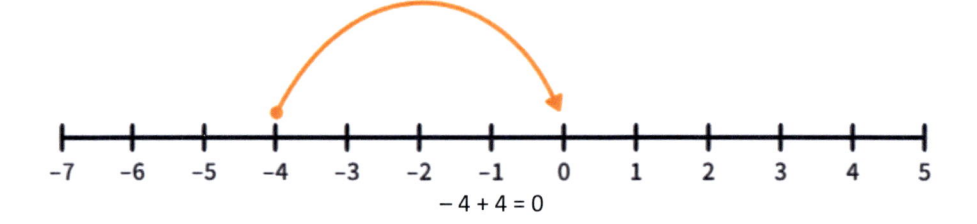

$-1 + 4 = 3$

$-2 + 4 = 2$

$-3 + 4 = 1$

$-4 + 4 = 0$

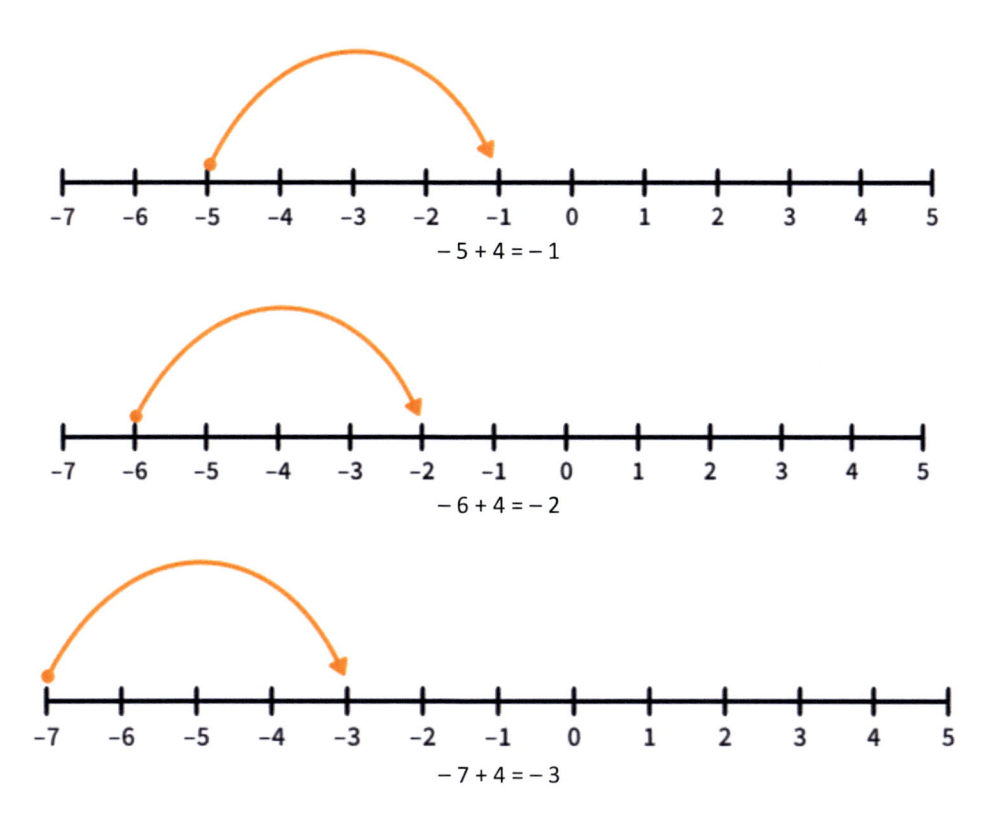

$$-5 + 4 = -1$$

$$-6 + 4 = -2$$

$$-7 + 4 = -3$$

These examples show that the statement is sometimes true. Can students articulate the condition under which the statement is always true?

Using Double-sided Counters

Here the counters have been assigned values of 1 and −1 and identified by their different colours. Also, the idea of zero pairs is employed as −1 + 1 = 0. Using this knowledge reveals the result of adding 4 to any negative number visible through the double-sided counter.

The diagrams below show the result as some values being positive numbers and others as negative numbers. Notice that there is no manipulative showing − 4 + 4 = 0 as it is difficult to show 'nothing' with the counters whereas this is visible on the number line shown above.

When students work with the idea of zero pairs with the counters, they are physically able to remove them to see the result. The arrow in the diagram directs to the result (sum) from the action of removing zero pairs.

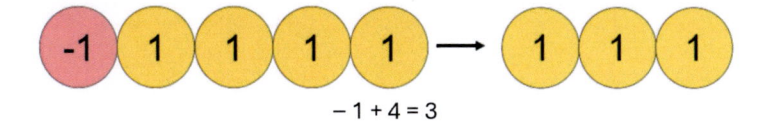

$$-1 + 4 = 3$$

$$-2 + 4 = 2$$

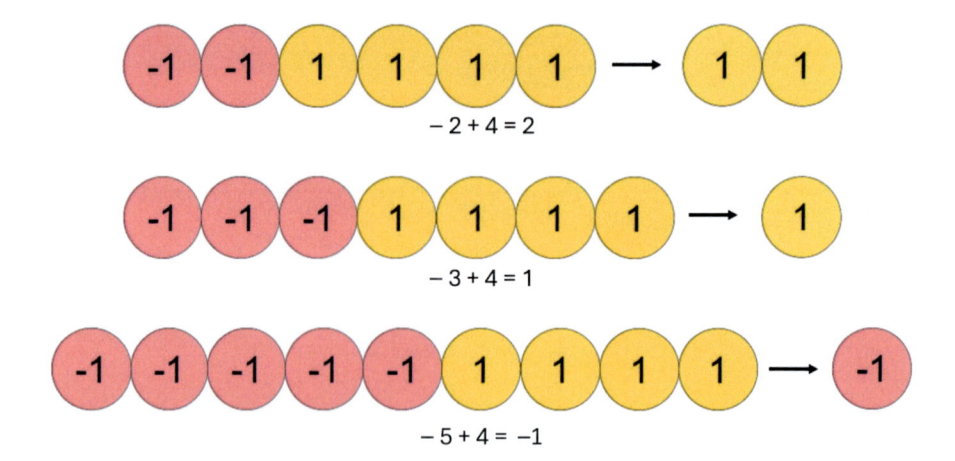

$$-3 + 4 = 1$$

$$-5 + 4 = -1$$

Again, the outcome from the exploration is the same. The statement is sometimes true.

Notes: Double-sided Counters used in the diagrams are from MathsBot.com and the Number Lines used in the diagrams are from Polypad.org.

The idea of zero-pairs is generally introduced in secondary. However, in primary, the concept of subtracting the same amount and using the commutative law supports this thinking when adding negative numbers.

Chapter 2

Multiplication and Division Statements

The number 20 can be arranged as an array in only two different ways. MD001	Numbers (integers) that end in a 4 are multiples of 4. MD008
The product of an odd number and an even number is an odd number. MD002	To divide a number by 10, just remove the last digit. MD009
The product of three numbers is the same as the sum of them. MD003	A number that is divisible by 3 is also divisible by 6. MD010
Doubling a number makes it bigger. MD004	Multiples of 8 are also multiples of 6. MD011
Halving a 2-digit number will result in a single digit. MD005	When a whole number is multiplied by 9, the sum of the digits of the product is 9. MD012
All numbers that end in zero (0) are multiples of twenty (20). MD006	The product of the diagonals of any 2x2 square on a multiplication grid are always equal (e.g. 6x12 and 9x8). MD013
Doubling a multiple of 5 will give a multiple of 10. MD007	Dividing a number by 1 will result in 1. MD014

When a whole number is divided by 5 the biggest remainder there can be is 4. MD015	The square of a number is greater than the number doubled. MD021
An even number divided by an odd number gives a quotient with no remainder. MD016	A square number always has an odd number of factors. MD022
When multiplying two whole numbers, the resulting product will be greater than either of the chosen numbers. MD017	Multiples of 6 are one more and one less than prime numbers. MD023
Halving a 3-digit number less than 200 gives a 2-digit answer. MD018	Dividing a number makes the quotient smaller than the dividend. MD024
Finding a third of a number is the same as dividing it by 3. MD019	Any whole number can be multiplied by partitioning (e.g. 35x4 is 30x4 plus 5x4). MD025
Finding a quarter of a number is the same as halving and halving again. MD020	

> **MD001 Statement:** The number 20 can be arranged as an array in only two different ways.
>
> **Answer:** NEVER TRUE
>
> **Manipulatives:** Cuisenaire Rods

This statement can be shown as never true through the use of *Cuisenaire rods* by physically showing the arrays using different rods of equal length/size. Encourage students to show more than two examples and discuss their findings alongside the relevance of the orientations where the example appears to look the same. In turn, this will provide the opportunity to discuss the commutative property of multiplication.

Examples of the arrangement of Cuisenaire rods is shown in the diagram below through the use of unit value of 1. Note that, here the assumption is that 20 rods of value 1 have been counted accurately and used to rearrange and create the arrays shown. Discussions that relate to the dimensions of the arrays will support the connection to times table knowledge.

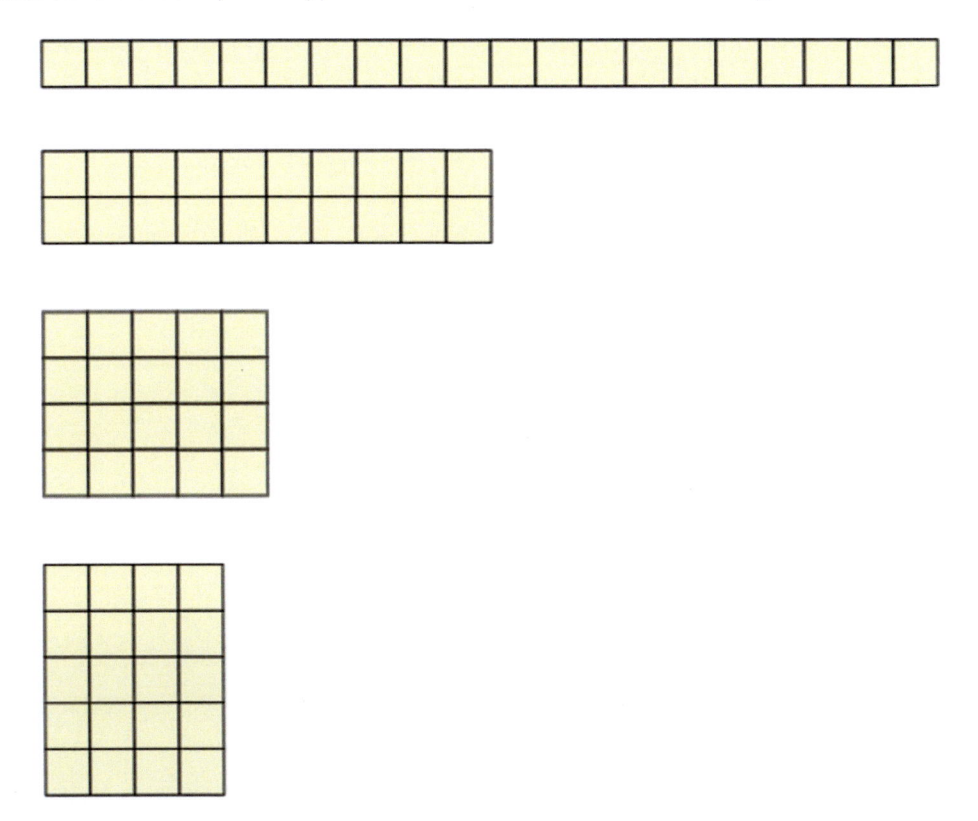

The example on the following page shows the use of different rods with a 2 by 10 and 10 by 2 arrangement when compared alongside the use of unit cubes. The different orientations expose the multiplication that array represents.

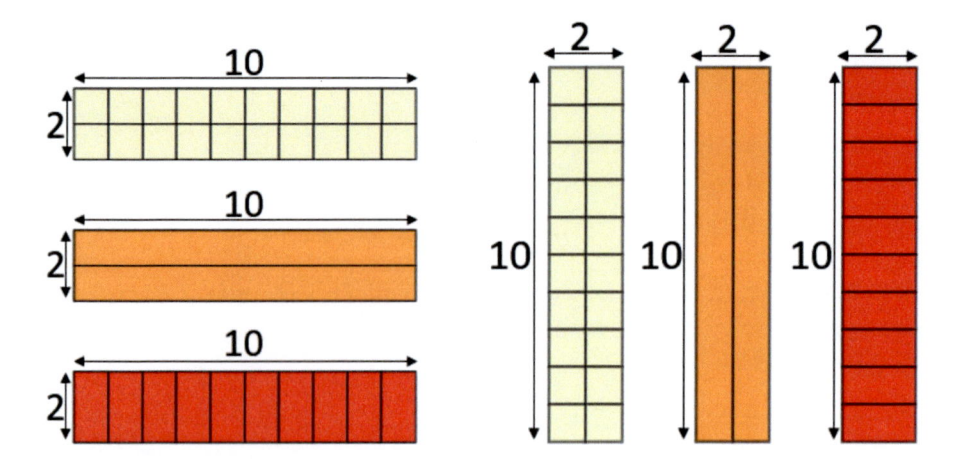

From the examples, the statement is never true.

> **Notes:** Cuisenaire Rods used in the examples are from MathsBot.com.
>
> Students may use other arrangements of the Cuisenaire rods instead of those shown in the examples above depending on previous knowledge of known facts.
>
> Other manipulatives such as maths link cubes or unit Dienes may be used. However, be aware that there could be a tendency to count the cubes individually.
>
> Students may approach the arrangement of the cubes sequentially by starting with a horizontal line of 20 cubes as seen in the example on the previous page. Then, progress to see what happens with the cubes if arranged in groups of 2, 3 etc. Discussions around divisibility or factors can arise from this path of investigation. Again, providing opportunities to articulate their findings using mathematical language.

> **MD002 Statement:** The product of an odd number and an even number is an odd number.
>
> **Answer:** NEVER TRUE
>
> **Manipulatives:** Cuisenaire Rods

This statement can be shown as never true through using *Cuisenaire rods* by identifying the value of the product using the area model. Encourage students to identify the product as the area of the rectangle created by the two side lengths as the odd number and even number. In the diagrams below, the side length is assumed to be of length 1 for every unit square.

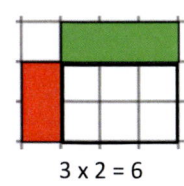

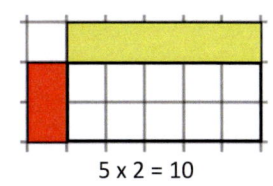

$$1 \times 2 = 2 \qquad 3 \times 2 = 6 \qquad 5 \times 2 = 10$$

In the number sentences above, students may start to conjecture that an odd number multiplied by an even number is equal to an even number. Encourage students to consider further examples and generalise.

The examples above reveal the statement to be never true.

> **Notes:** Cuisenaire Rods used in the examples are from MathsBot.com.
>
> Multiplication grids and number lines can be used as alternative manipulatives.
>
> Students may consider commutativity and interpret the original statement as: "The product of and even number and an odd number is an odd number".
>
> Students may also like to investigate other statements such as: "The product of two even numbers is even" or "The product of two odd numbers is odd".
>
> Furthermore, students may discuss the case when any number is multiplied by zero. For example, by considering the definition of an even number – a number is said to be even if it is divisible by 2 and it is an integer.
> $\frac{0}{2} = 0$ so, $0 = 2 \times 0$.
>
> Alternatively, considering all multiples of 2 are even numbers, we can say that $2 \times m$ is an even number, where m is a positive integer.
>
> Using the same consideration, we can say that an odd number is defined as one more than an even number. In which case, $2 \times m + 1$ is an odd number, where m is a positive integer. So, if 0 is an odd number then $2m + 1 = 0$. However, this equates m to hold a value of $-\frac{1}{2}$, which is not a positive integer. Thus, showing that 0 is not an odd number.

MD003 Statement: The product of three numbers is the same as the sum of them.

Answer: SOMETIMES TRUE

Manipulatives: Dienes

This statement can be shown as sometimes true using *Dienes* by creating the models that represent the products of three chosen numbers and similarly for the sum of the same three numbers. Here it is important to distinguish and be explicit in the discussions about the representation of products in terms of dimensions.

The example below shows the statement to be true for the chosen number values. Encourage students to explore any other cases where the statement is true. The assumptions made here are that the students can create models that represent multiplication as well as addition using the same manipulative.

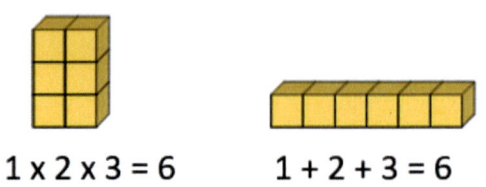

1 x 2 x 3 = 6   1 + 2 + 3 = 6

Whereas the example below shows the statement to be not true for the chosen number values. Encourage students to find other cases where the statement is not true.

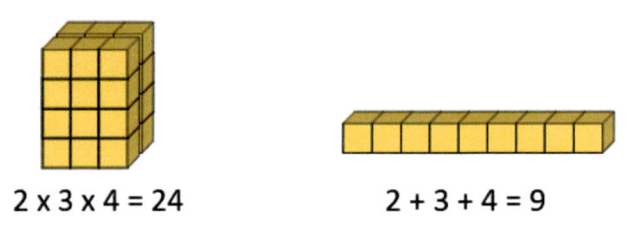

2 x 3 x 4 = 24   2 + 3 + 4 = 9

From the examples shown, the statement is sometimes true.

Notes: Dienes used in the examples are from MathsBot.com.

Maths link cubes are an alternative manipulative that can be used.

The statement does not specify if the three numbers have to be different or the same or a combination. As a result, encourage students to investigate patterns within one specific condition and make predictions.

Students who need scaffolding for multiplication will benefit from using a multiplication grid and a number line for addition.

Students may also investigate the statement for more than three numbers.

MD004 Statement: Doubling a number makes it bigger.

Answer: SOMETIMES TRUE

Manipulatives: Number Lines

This statement can be shown as sometimes true through the use of *Number Lines* by applying the idea that doubling implies two jumps of the same amount.

The example below shows the statement to be true for the chosen positive whole number. Encourage students to find other cases where the statement is true.

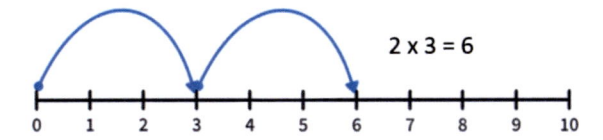

$2 \times 3 = 6$

Whereas the example below shows the statement to be not true for the chosen negative number. Encourage students to find other cases where the statement is not true.

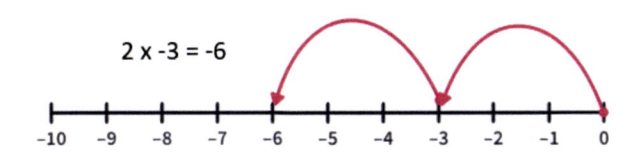

$2 \times -3 = -6$

From the above examples, the statement appears to be sometimes true.

Notes: Number Lines used in the examples are from Polypad.org.

The statement does not specify the type of number to be used. If students are only familiar with positive numbers, then they will reach the conclusion that the statement is **always true**. This may well be appropriate for them. However, for other students it may be appropriate to encourage them to think about negative numbers. Note that even if students have not been taught about multiplying negative numbers, the concept of doubling can be used to explore what happens with negative numbers. It can further be discussed by relating it to temperature to make the connection with size – the further away from zero (on the left of the number lines as shown in the examples above), the smaller the number.

Students could also be encouraged to consider decimals or fractions or numbers smaller than 1. They could investigate patterns within one of these specific conditions and make predictions/generalisations based on their choice.

The special case of zero as a digit is not considered here, particularly as multiplying by zero is difficult to present visually. Students may choose to make this a condition for their investigation.

MD005 Statement: Halving a 2-digit number will result in a single digit number.

Answer: SOMETIMES TRUE

Manipulatives: Cuisenaire Rods

This statement can be shown as sometimes true through using *Cuisenaire rods* by applying the idea of symmetry.

The example below shows the statement to be true for the number 14. Here, it can be seen that the number 14 has been created using rods of value 10 and 4. The diagram underneath shows that the 10 rod has been substituted with a 4 rod and two 3 rods purposely to make the symmetry visually explicit. The assumption is that students are able to identify combinations that will support the idea of symmetry in numbers as a connection with doubling and halving. Blue vertical markings are used in the diagrams below as a line of symmetry. Thus, resulting in the value of half of the original amount and in turn able to calculate the value of half of 14 to equal 7 (sum of 4 and 3).

Students may partition 14 in alternative ways and discuss whether that still results in the value of half of 14 being 7.

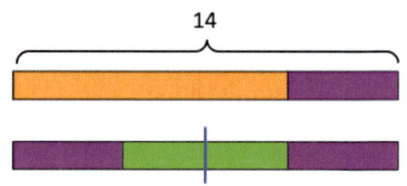

Encourage students to find other cases where the statement is true.

Be aware, in the case of using a 2-digit odd number, the Cuisenaire rods do show symmetry, however, creating markings on the unit rod maybe required to visually show the equality and symmetry. The concept of half could be explored through something that students can relate to, such as half of an apple, half of 2 apples, half of 3 apples and so forth.

The diagram below shows how this has been approached using mark making for the number 13. Again, here students can identify that the size of the bottom bars on the left-hand side to the blue vertical marking is the same as that on the right-hand side of the mark. Thus, revealing the 'half-ness' as well as the symmetry and in turn able to calculate the value of half of 13 to equal $6\frac{1}{2}$ (The sum of $3 + 3 + \frac{1}{2}$).

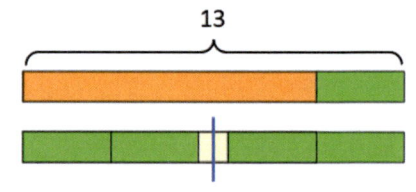

For the chosen number 26, the statement shows to be not true. The diagram on the following page, shows how the "half-ness" can be visualised. Therefore, half of 26 is equal to 13 (the sum of 10 and 3), a 2-digit number and not a single digit.

Encourage students to find other cases where the statement is not true and consider what assumptions and deductions they make from their findings.

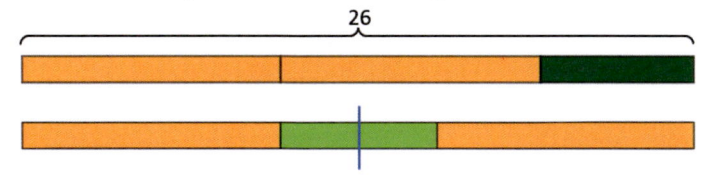

Students may choose to show another arrangement that reveals symmetry, for example placing bars one under the other as seen on the diagram on the right, instead of that seen above for half of 26. Note that this overlaps with the multiplicative concept rather than solely additive.

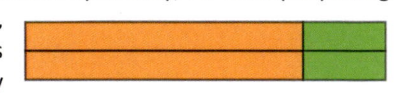

An example of a 2-digit odd number can be seen in the diagram below with the number 23, which reveals the value of half of 23 as $11\frac{1}{2}$ (the sum of $10 + 1 + \frac{1}{2}$).

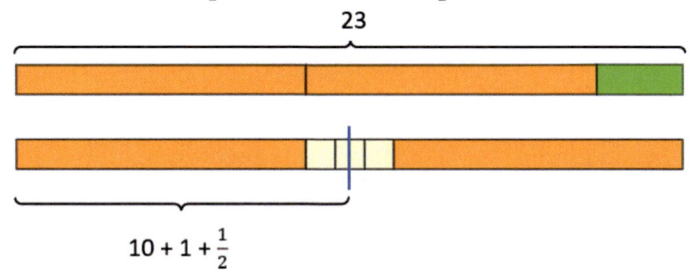

The examples above show the statement to be sometimes true.

> **Notes:** Cuisenaire Rods used in the examples are from MathsBot.com.
>
> Where students are unfamiliar with their number bonds or do not recognise that a number can be replaced or substituted by a combination of smaller numbers, then support by posing questions that navigate towards symmetry.
>
> In the first example with 14, if a student doesn't recognise and apply their knowledge of number bonds to 10, then pose the question – Can the 10 be replaced with a 4 and something? This idea employs symmetry by having a 4 rod on both sides of the symmetrical line.
>
> This approach leads to creating the model shown below, where the 10 rod is replaced by a 4 rod and a 6 rod:
>
> The 6 rod can be replaced with two 3 rods, as seen here. Hence revealing the "half-ness". 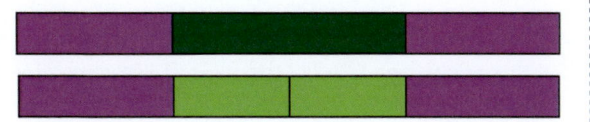
>
> The assumption made here is that students would be familiar with Cuisenaire rods and that the smallest unit rod is of value 1. The rest of the rods are proportionate in size. Refer to the image provided in the introduction page for values associated with the rods.

MD006 Statement: All numbers that end in zero (0) are multiples of twenty (20).

Answer: SOMETIMES TRUE

Manipulatives: Cuisenaire Rods

This statement can be shown as sometimes true through the use of *Cuisenaire rods* by placing the rods in groups of 20. This can be achieved by placing two 10 rods as seen on the furthest left image in the diagram below.

Here we have chosen to investigate the numbers 10, 20, 30, 40, 50, 60 and 70. By placing the rods in groups of 20, it can be seen that there is a pattern emerging.

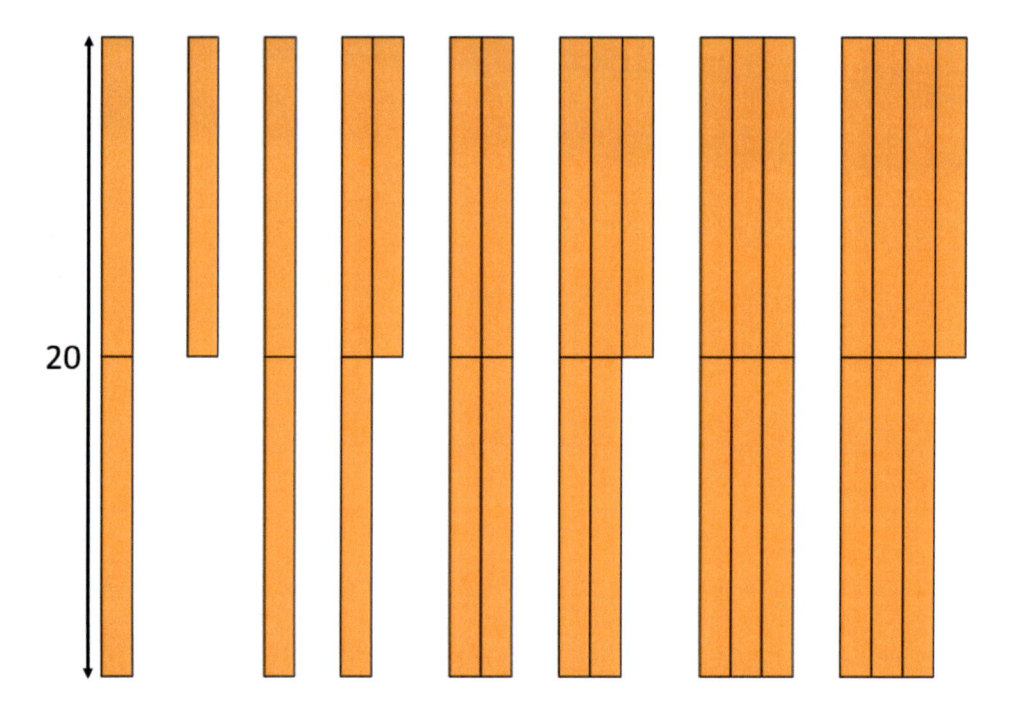

The numbers 20, 40 and 60 (as shown above) are multiples of 20, whereas the others (10, 30, 50 and 70) are not – thus making the statement sometimes true.

Encourage the students to make predictions and generalisations that lead to identifying that the number ending in zero cannot just be divisible by 2 and 10 in order for it to be divisible by 20. For example, 30 is divisible by 2 as well as 10, however 30 ÷ 20 yields a non-whole number quotient. Therefore, when making generalisations, students need to consider the conditions under which the generalisations are valid.

A form of scaffold can be to encourage students to write the size of the number that is represented by the rods next to each different representation rather than reverting to counting in 10s to support their mathematical thinking.

Notes: Cuisenaire Rods used in the examples are from MathsBot.com.

An assumption made in the examples above is that students understand the idea of grouping.

Further investigations can be embarked upon by predicting whether the same approach can be applied by altering the statement to: 'All numbers that end in 00 are multiples of 200', and/or consider the concept of grouping and how it relates to the factors of a given number.

The first 20 multiples of 20 are: 20, 40, 60, 80, 100, 120, 140, 160, 180, 200, 220, 240, 260, 280, 300, 320, 340, 360, 380 and 400.

> **MD007 Statement:** Doubling a multiple of 5 will give a multiple of 10.
>
> **Answer:** ALWAYS TRUE
>
> **Manipulatives:** Cuisenaire Rods

This statement can be shown as always true through the use of *Cuisenaire rods* by placing focus on the 5 rod as the constant and presenting it as seen in the diagram below.

The rods have been specifically arranged in this manner to support the connection with the area model. In the diagrams, the green arrows show the physical transition once the action of doubling has taken place. The dimensions are labelled for emphasis on conversion. Encourage the students to discuss the connections they make with the dimensions including their jottings and their choice of methods.

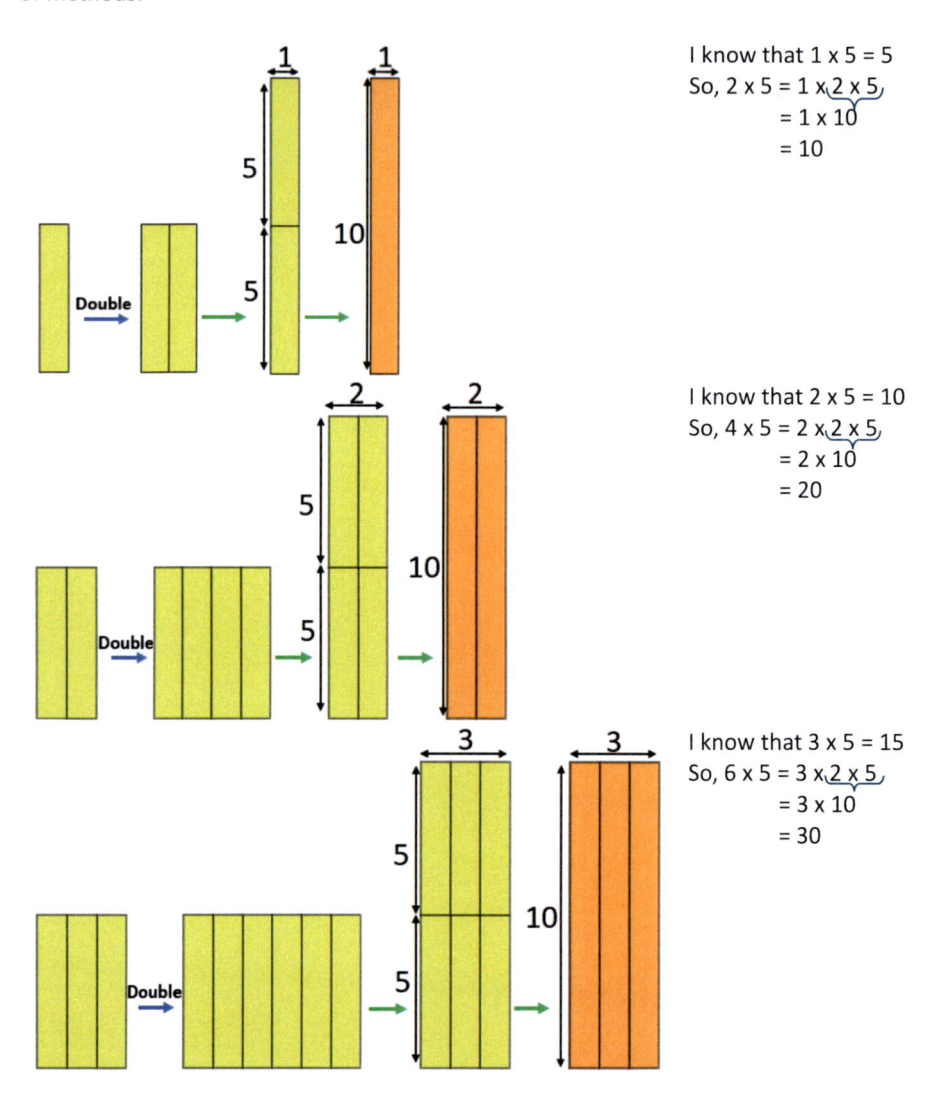

I know that 1 x 5 = 5
So, 2 x 5 = 1 x 2 x 5
  = 1 x 10
  = 10

I know that 2 x 5 = 10
So, 4 x 5 = 2 x 2 x 5
  = 2 x 10
  = 20

I know that 3 x 5 = 15
So, 6 x 5 = 3 x 2 x 5
  = 3 x 10
  = 30

Students may show other forms of numerical methods to identify the value of the product. For example:
Double 5 = 5 + 5 = 10
Double 10 = 10 + 10 = 20
Double 15 = 15 + 15 = 30 etc.

From the examples shown the statement is always true.

> **Notes:** Cuisenaire Rods used in the examples are from MathsBot.com.
>
> An assumption made in the examples shown above is that students understand the idea of doubling and are able to calculate the doubled values.
>
> An alternative resource that can be used is the multiplication grid, by obscuring the other multiplication facts as seen in the diagram here.
>
> Further predictions involving the doubling relationship with other multiplication facts can be made and investigated.
>
> The first 20 multiples of 5 are: 5, 10, 15, 20, 25, 30, 35, 40, 45, 50, 55, 60, 65, 70, 75, 80, 85, 90, 95 and 100.
> The first 20 multiples of 10 are: 10, 20, 30, 40, 50, 60, 70, 80, 90, 100, 110, 120, 130, 140, 150, 160, 170, 180, 190 and 200.

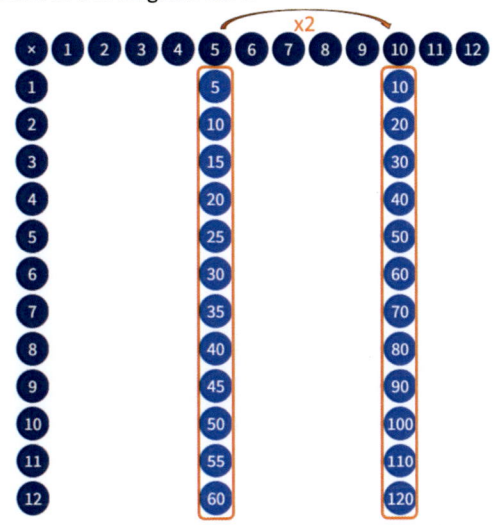

MD008 Statement: Numbers (integers) that end in a 4 are multiples of 4.

Answer: SOMETIMES TRUE

Manipulatives: Number Tiles

This statement can be shown as sometimes true through the use of *Number Tiles* by placing focus on the fact that all multiples of 4 can be arranged in groups of 4.

The tiles have been specifically arranged in this manner to support the connection with the area model representation of a given number. In the diagrams below, it can be seen that we have used a systematic approach and made jottings of the numbers alongside it's representation with the tiles.

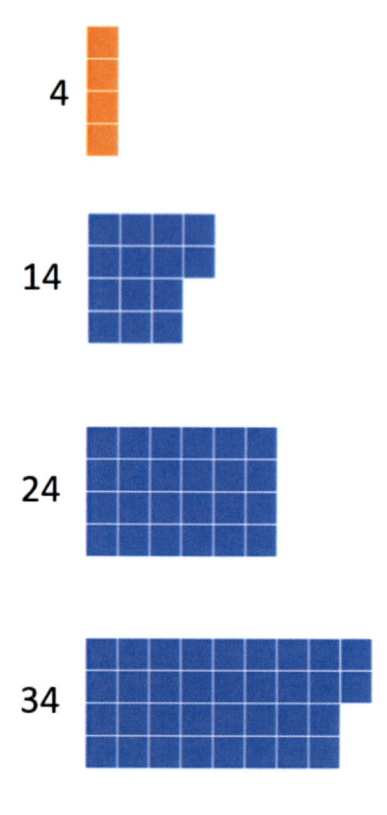

4

14

24

34

To promote the deductions and predictions from pattern spotting, encourage students to discuss their findings as they group the tiles in fours. This links well with factors and the idea of remainders. Students may infer from the above examples that the numbers 4 and 24 are multiples of 4 as they show one complete rectangle with one of its dimensions as 4. In this case, encourage the students to investigate further with 44, 54, 64, 74 to reveal a pattern that leads to considering conditions under which this statement is true or not true.

Other resources such as number grids can be used to emphasize a pattern, as shown in the diagram on the following page, where the emphasis is on the numbers that validate the statement to be true.

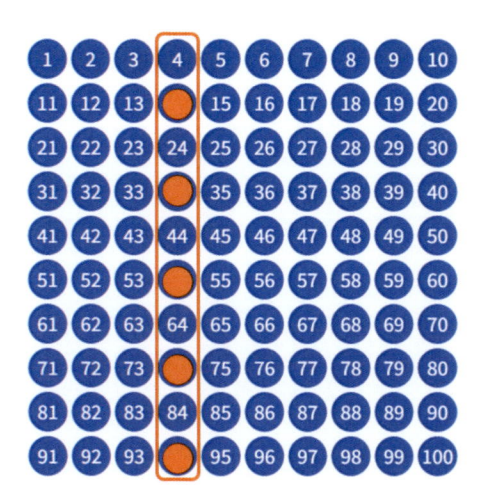

Students may record their jottings in a systematic manner to reveal any patterns like this:

4 ÷ 4 = 1

14 ÷ 4 = 3r2

24 ÷ 4 = 6

34 ÷ 4 = 8r2

44 ÷ 4 = 11

54 ÷ 4 = 13r2

64 ÷ 4 = 16

74 ÷ 4 = 18r2

84 ÷ 4 = 21

From the examples shown the statement is sometimes true.

> **Notes:** Number Tiles and Number Grid used in the examples are from Polypad.org
>
> Students may wish to consider negative integers as well as positive integers.
>
> Students may also investigate further similar patterns from the number grid.

MD009 Statement: To divide a number by 10, just remove the last digit.

Answer: SOMETIMES TRUE

Manipulatives: Number Grid, Dienes

This statement can be shown as sometimes true through the use of a *Number Grid* with jottings and *Dienes*.

Be aware that many students may generalise about dividing a number by 10 and assume that they can just remove the last digit as a maths trick. This misconception mainly occurs when students have only worked with multiples of 10. Therefore, students should be encouraged to work with a variety of numbers. Ensure the focus is on place value and how the digits are moved one place to the right in a division by 10.

Using Number Grid

In the diagrams below, the act of removing the last digit is indicated by using the orange strokes.

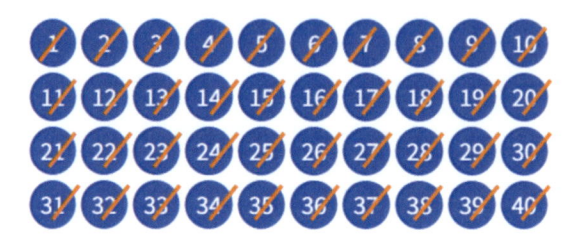

Alongside the markings on the grid, jottings are made to reflect the interpretation of the marking in relation to the statement. For example:

For the first row: 1 ÷ 10 = 0, 2 ÷ 10 = 0, 3 ÷ 10 = 0, 4 ÷ 10 = 0 …… 10 ÷ 10 = 1
The above strategy results in all the single digit numbers being 0 as a representation of nothing in the ones place. However, 1 ÷ 10 = 0.1, 2 ÷ 10 = 0.2, 3 ÷ 10 = 0.3, 4 ÷ 10 = 0.4 etc. and therefore not 0.

For the second row: 11 ÷ 10 = 1, 12 ÷ 10 = 1, 13 ÷ 10 = 1, 14 ÷ 10 = 1 …… 20 ÷ 10 = 2
Again, the above strategy results in all the digits in the ones place being removed. However, 11 ÷ 10 = 1.1, 12 ÷ 10 = 1.2, 13 ÷ 10 = 1.3, 14 ÷ 10 = 1.4 etc. and therefore the quotient is greater than 1 and not just equal to 1.

The jottings above can be used to identify patterns. Encourage students to specify conditions under which the statement could be true and by what means they would justify their findings. Can they articulate that the statement is true for multiples of 10 only?

Using Dienes

The diagrams on the following page show clearly that the statement is true for the number 110 however, not true for 111.

110 ÷ 10 = 11

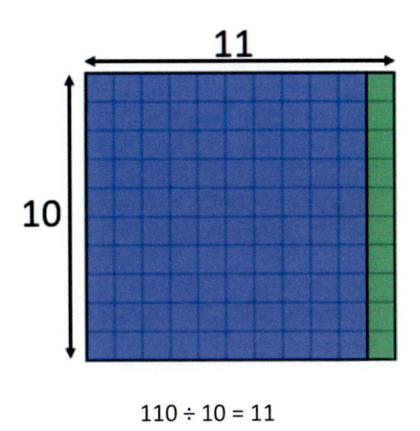

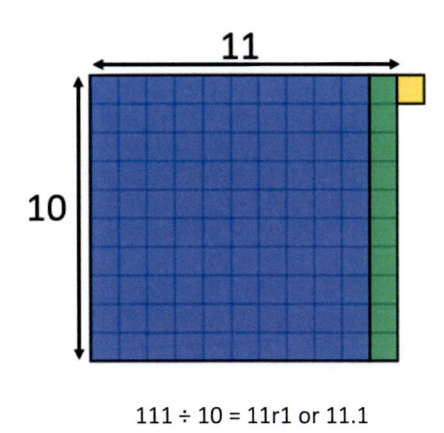

111 ÷ 10 = 11r1 or 11.1

From the examples shown the statement is sometimes true.

Notes: Number Grid used in the example is from Polypad.org and Dienes used in the example is from MathsBot.com.

Students may wish to investigate further similar patterns with other types of numbers e.g.: decimals and fractions or other powers of 10, for example, the effect of dividing by 100 or 1000.

The statement could be rephrased as "To divide a number by 10, just remove the last digit from the dividend."

Remember: Dividend ÷ Divisor = Quotient

MD010 Statement: A number that is divisible by 3 is also divisible by 6.

Answer: SOMETIMES TRUE

Manipulatives: Cuisenaire Rods

This statement can be shown as sometimes true through the use of a *Cuisenaire rods* and jottings.

Assumptions made here are that the students understand the definition of an even number along with the concept of division. Here we look at division through equal grouping using Cuisenaire rods. In the diagrams below, the arrows indicate the physical movement of the rods, whilst the blue lines on the edge of the rods are replicating the division bracket such that the total value of the rods is the dividend. This lands itself neatly with the area model as it creates a visual relationship between multiplication and division.

Note that students may wish to refer to a multiplication grid alongside the Cuisenaire rods.

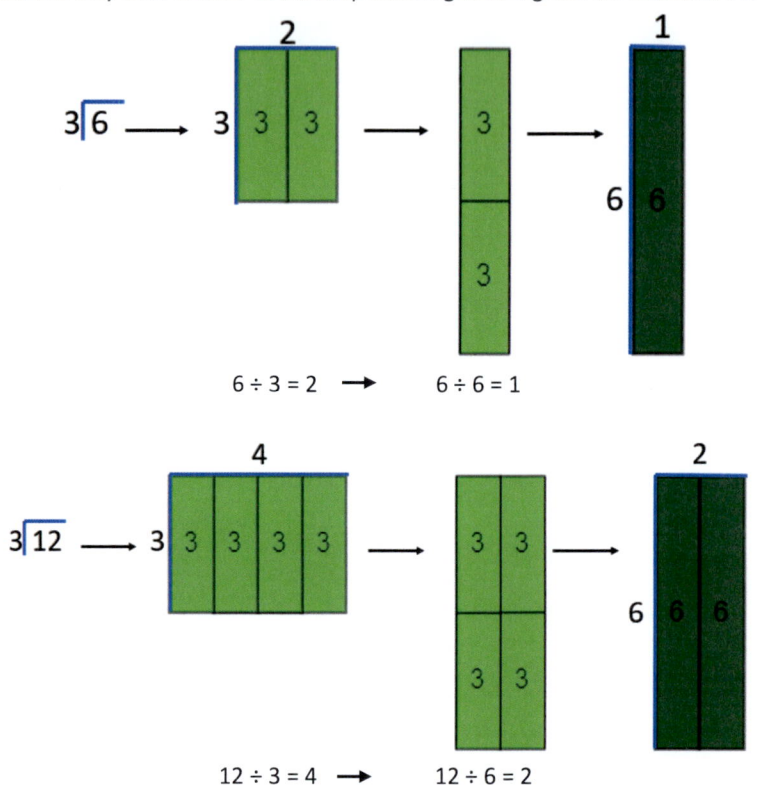

$6 \div 3 = 2 \rightarrow 6 \div 6 = 1$

$12 \div 3 = 4 \rightarrow 12 \div 6 = 2$

Jottings as in the examples on the following page, support pattern spotting discussions. Encourage students to make predictions from the relationships they identify and convince someone else about their discoveries.

$6 \div 3 = 2$ $\quad\quad$ $12 \div 3 = 4$ $\quad\quad$ $18 \div 3 = 6$

x2 ↓ ↓ ÷2 $\quad\quad$ x2 ↓ ↓ ÷2 $\quad\quad$ x2 ↓ ↓ ÷2

$6 \div 6 = 1$ $\quad\quad$ $12 \div 6 = 2$ $\quad\quad$ $18 \div 6 = 3$

So far, the examples above show the statement to be true.

However, the following example considers another multiple of 3, which shows the statement to be not true.

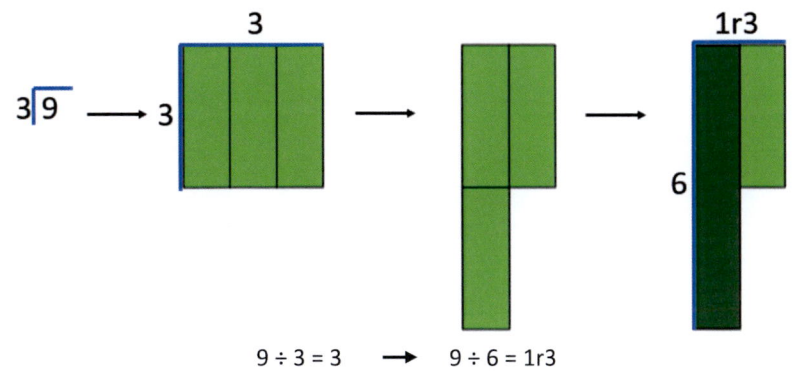

$9 \div 3 = 3$ → $9 \div 6 = 1r3$

Hence, from the examples shown the statement is sometimes true.

> **Notes:** Cuisenaire Rods used in the examples are from MathsBot.com.
>
> Students may use other orientations with the Cuisenaire rods, in which case monitor consistency in their approach to support pattern spotting and conjecturing.
>
> Further investigations with doubling and halving may be undertaken with other numbers to form a generalisation.
>
> First 10 multiples of 3 are: 3, 6, 9, 12, 15, 18, 21, 24, 27, 30.
> First 10 multiples of 6 are: 6, 12, 18, 24, 30, 36, 42, 48, 54, 60.
>
> Notice that we have listed the first 10 'multiples of' instead of 'divisible by' as 0 can be divided by both 3 and 6. Thus, leading to an opportunity to discuss multiplicative reasoning.

MD011 Statement: Multiples of 8 are also multiples of 6.

Answer: SOMETIMES TRUE

Manipulatives: Cuisenaire Rods

This statement can be shown as sometimes true through the use of *Cuisenaire rods and jottings*. In the diagrams below, the rods have not been labelled with a number value (refer to the introductory page for the default value of the rods). The idea of the area model has been used to compare and convert between the 'group of 8' to 'group of 6'. The arrows indicate the movement and replacement of the 8 rod to show how it can be substituted with a 6 rod.

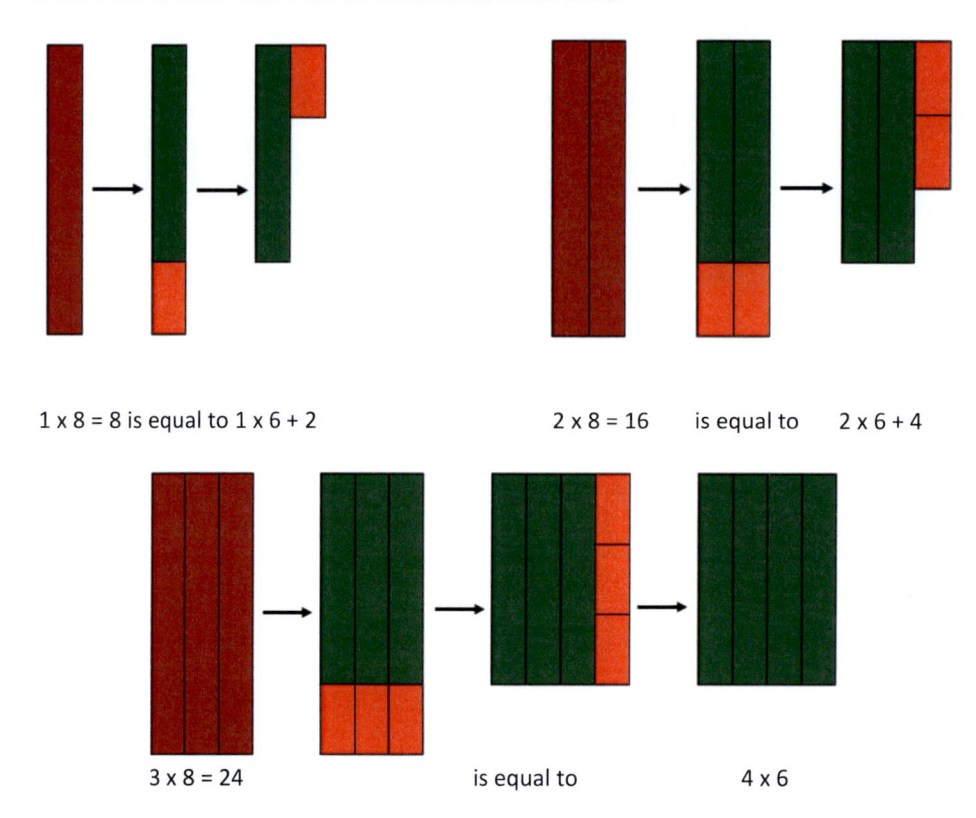

1 x 8 = 8 is equal to 1 x 6 + 2   2 x 8 = 16   is equal to   2 x 6 + 4

3 x 8 = 24   is equal to   4 x 6

In the example above, 8 and 16 can be shown to be multiples of 8 but not of 6, leading students to believe the statement is not true. However, the statement is true for 24 which is shown to be a multiple of both 6 and 8.

As students continue to investigate, encourage identifying a pattern. They may notice a pattern emerging with the factors of 24. Leading to the generalisation that a number can be a multiple of 6 and 8 if and only if it is an even number and divisible by 3 as well as divisible by 8. A way of presenting their work can be seen in the diagrams on the following page.

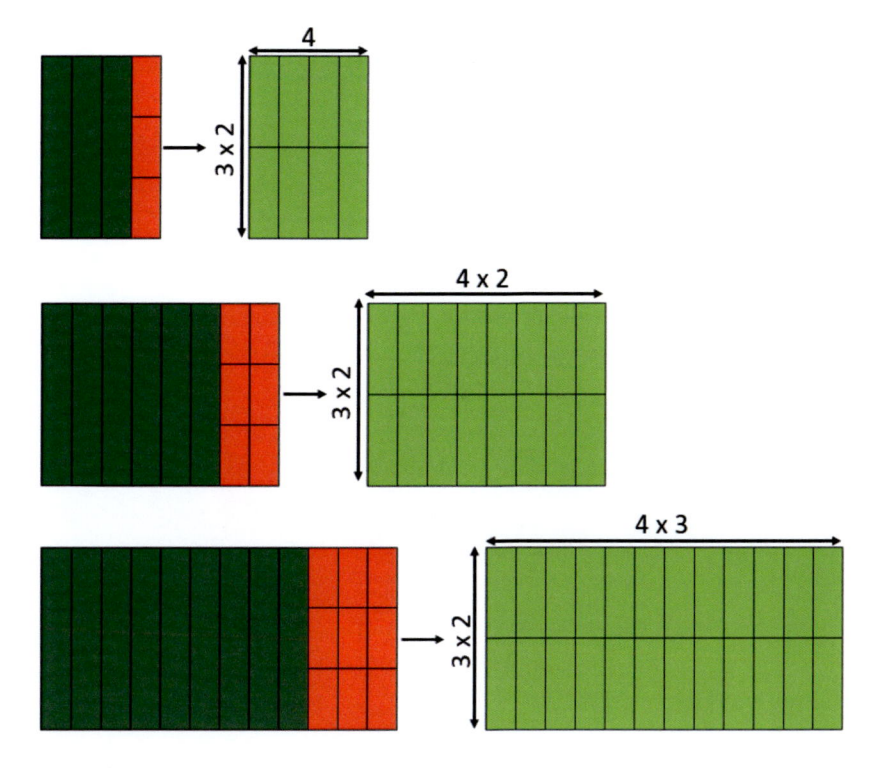

From the examples shown the statement is sometimes true.

> **Notes:** Cuisenaire Rods used in the examples are from MathsBot.com.
>
> Students may use a multiplication grid to compare and identify common multiples. In which case, encourage to search for a reason as to why there is a commonality with multiples of 24.
>
> Further investigations with similar connections between other multiples or divisibility rules may be undertaken.
>
> Divisibility rule of 6 – A number is divisible by 6 if it is an even number and is divisible by 3.
>
> Divisibility rule of 8 – A number is divisible by 8 if the last three digits of the number are divisible by 8.
>
> First 20 multiples of 6: 6, 12, 18, 24, 30, 36, 42, 48, 54, 60, 66, 72, 78, 84, 90, 96, 102, 108, 114, 120.
>
> First 20 multiples of 8: 8, 16, 24, 32, 40, 48, 56, 64, 72, 80, 88, 96, 104, 112, 120, 128, 136, 144, 152, 160.

MD012 Statement: When a whole number is multiplied by 9, the sum of the digits of the product is 9.

Answer: ALWAYS TRUE

Manipulatives: Cuisenaire Rods

This statement can be shown as always true through the use of *Cuisenaire rods.* In the diagrams below, the rods have not been labelled with a number value (refer to the introductory page for the default value of the rods). The arrows are used to indicate the movement as we compare the 9 rod with the 10 rod to understand the appearance of the value 9. As the base 10 system is used, we can notice that in the tens place the tens value increases by 1 and the ones place decreases by 1. This pattern is observed in the diagrams below, hence, maintaining the value of 9.

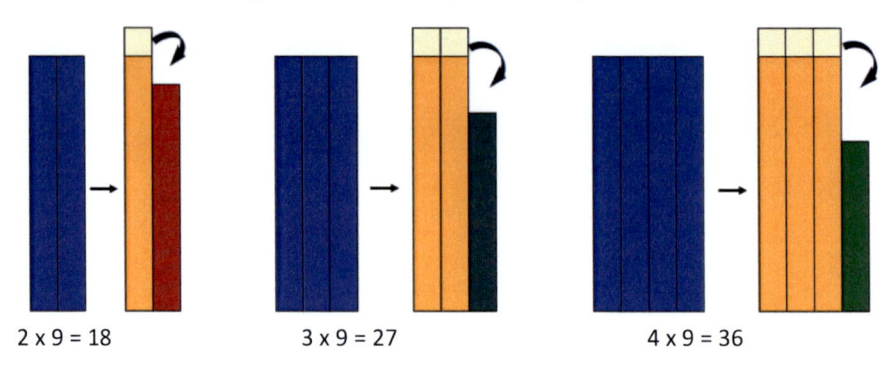

2 x 9 = 18   3 x 9 = 27   4 x 9 = 36

Multiples of 9	Sum of the digits of the product
1 x 9 = 9	9
2 x 9 = 18	1 + 8 = 9
3 x 9 = 27	2 + 7 = 9
4 x 9 = 36	3 + 6 = 9
5 x 9 = 45	4 + 5 = 9
⋮	⋮
11 x 9 = 99	9 + 9 = 18 → 1 + 8 = 9
12 x 9 = 108	1 + 0 + 8 = 9
13 x 9 = 117	1 + 1 + 7 = 9
14 x 9 = 126	1 + 2 + 6 = 9 etc

A table of values can be used to notice the increasing and decreasing patterns of the digits of the products. Every time 9 is added to find the next multiple from the previous multiple, we increase the place value in the tens place value by one and reduce the ones place value by one. Hence, the sum of 9 is revealed for every multiple of 9.

From the examples shown the statement is always true.

Notes: Cuisenaire Rods used in the example are from MathsBot.com.

Students can use manipulatives such as number lines, maths link cubes or number frames to support adding, however, where possible encourage the use of their known number facts to mentally compute the sum of the digits of the multiple of 9.

Further investigations that present a similar approach such as the multiples of 11 may be investigated or other multiples such as "All multiples of 8 are also multiples of 4".

> **MD013 Statement:** The product of the diagonals on any 2x2 square on a multiplication grid are always equal (e.g. 6 x 12 and 9 x 8).
>
> **Answer:** ALWAYS TRUE
>
> **Manipulatives:** Multiplication Grid

This statement can be shown as always true by investigating 2 by 2 squares on a *Multiplication Grid*.

Encourage students to select their own 2 by 2 box and explore the products, as well as use a range of manipulatives to check their multiplications.

In the diagram on the right, the 2 by 2 square box chosen would give diagonal products of 72:
6 x 12 = 72 and 8 x 9 = 72.

×	1	2	3	4	5	6	7	8	9	10	11	12
1	1	2	3	4	5	6	7	8	9	10	11	12
2	2	4	6	8	10	12	14	16	18	20	22	24
3	3	6	9	12	15	18	21	24	27	30	33	36
4	4	8	12	16	20	24	28	32	36	40	44	48
5	5	10	15	20	25	30	35	40	45	50	55	60
6	6	12	18	24	30	36	42	48	54	60	66	72
7	7	14	21	28	35	42	49	56	63	70	77	84
8	8	16	24	32	40	48	56	64	72	80	88	96
9	9	18	27	36	45	54	63	72	81	90	99	108
10	10	20	30	40	50	60	70	80	90	100	110	120
11	11	22	33	44	55	66	77	88	99	110	121	132
12	12	24	36	48	60	72	84	96	108	120	132	144

This diagram shows four further examples – each chosen to be in the corner of the grid to check if this makes a difference.

From the top and working from left to right, the products of the diagonals are:

1 x 4 = 4 and 2 x 2 = 4, here we can see the products are the same.

11 x 24 = 264 and 22 x 12 = 264, here we can see the products are the same.

24 x 11 = 264 and 12 x 22 = 264, again, the products are the same and indeed this box is the same as the one before it.

×	1	2	3	4	5	6	7	8	9	10	11	12
1	1	2	3	4	5	6	7	8	9	10	11	12
2	2	4	6	8	10	12	14	16	18	20	22	24
3	3	6	9	12	15	18	21	24	27	30	33	36
4	4	8	12	16	20	24	28	32	36	40	44	48
5	5	10	15	20	25	30	35	40	45	50	55	60
6	6	12	18	24	30	36	42	48	54	60	66	72
7	7	14	21	28	35	42	49	56	63	70	77	84
8	8	16	24	32	40	48	56	64	72	80	88	96
9	9	18	27	36	45	54	63	72	81	90	99	108
10	10	20	30	40	50	60	70	80	90	100	110	120
11	11	22	33	44	55	66	77	88	99	110	121	132
12	12	24	36	48	60	72	84	96	108	120	132	144

121 x 144 = 17424 and 132 x 132 = 17424, again, the products are the same.

Students could use algebra if this was deemed appropriate. The diagram on the right shows how the algebra might work.

The rows have been labelled as a and the columns as b. So, the relationship between a and b has been given in an example 2 by 2 box.

	b				
×	1	2	3	4	5
1					
2			ab	a(b+1)	
3			(a+1)b	(a+1)(b+1)	
4					
5					

(rows labelled a)

Multiplying the first diagonal would give:

a b x (a+1) (b + 1)

= ab x (ab + a + b + 1)

= $a^2 b^2 + a^2 b + a b^2 + a b$

Multiplying the second diagonal would give:

a (b + 1) x (a+1) b

= (ab + a) x (ab + b)

= $a^2 b^2 + a^2 b + a b^2 + a b$

Here the brackets have been used for purpose of clarity with the expansions of them. Comparing the expressions for the products of both diagonals, we can state that they are equal.

From the examples shown the statement is always true.

> **Notes:** Multiplication grids used in these images are from MathsBot.com.
>
> An opportunity to explore commutativity with multiplication can be undertaken with this exercise.
> Students may further predict what happens with a 3 by 3 box or other square sized boxes as well as a 2 by 2.
>
> Furthermore, the multiplication grids can include other numbers such as, zero, negatives, fractions, decimals etc.

MD014 Statement: Dividing a number by 1 will result in 1.

Answer: SOMETIMES TRUE

Manipulatives: Counters, Number Lines

Using Counters

This statement can be shown as sometimes true through using *Counters* and considering the wording of dividing by one.

First of all, consider cases where it is not true.
If we have 4 counters and 'divide by 1' or give them to 1 person – how many counters does that person get? The person gets 4.
Mathematically: $4 \div 1 = 4$.

5 flowers in one group could be shown as 5 flowers in one vase, where the vase represents the group. $5 \div 1 = 5$ flowers.

And with counters we can see that this might be represented as:

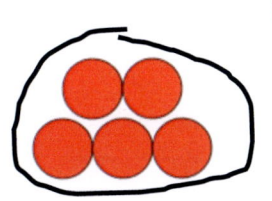

Diagrams that resemble the counters like dots or circles on square paper can be drawn instead, as well as, writing number sentences to support the findings.

Now let's consider when the statement is true:

In the case of one counter being in one group, $1 \div 1 = 1$.
Therefore, the calculation is an example of dividing by 1 where the answer is 1.

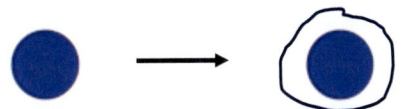

Using Number Lines

4 children grouped into one group can be modelled as beginning on 4 on a number line and then jumping back four (as in one group of 4):

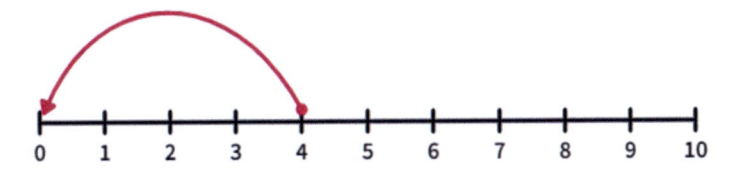

This shows 4 ÷ 1 = 4 (an example where the statement is not true).

Encourage students to explore other numbers on the number line.

From the examples shown the statement is sometimes true, particularly because of the special case where 1 ÷ 1 = 1.

> **Notes:** Counters used in the examples are from MathsBot.com and the Number Lines in the examples are from Polypad.org.
>
> This is a good opportunity to explore quotative (grouping or subtracting) and partitive division (sharing). It is important to be able to visualise the 'groups of one' structure as well as that of 'one group', to support the understanding of the two structures of division.
>
> For example: 3 children (represented by counters) divided into groups of 1 gives one child in each group and we can write 3 ÷ 3 = 1
>
> Similarly with the example with the number line:
> While four children divided into groups of one, would give four groups, shown by the four jumps below.
> 4 ÷ 4 = 1 showing the statement to be not true.
> (Also, we can say there are four groups of 1, there are four altogether. 4 x 1 = 4). 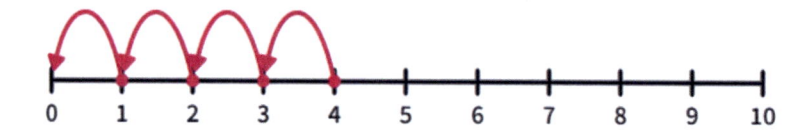
>
> Students may also state the fact that a number divided by 1 will give the same number they began with. The dividend and quotient will be equal when the divisor is 1.
>
> Students may further investigate what happens when they multiply a number by 1 or by 0.
>
> dividend ÷ divisor = quotient
> dividend ÷ 1 = divisor

> **MD015 Statement:** When a whole number is divided by 5 the biggest remainder there can be is 4.
>
> **Answer:** ALWAYS TRUE
>
> **Manipulatives:** Counters

This statement can be shown as always true through using *Counters* and trying different numbers divided by 5.

First of all, let's pick a multiple of 5, say 10:

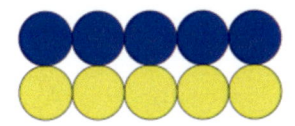

10 ÷ 5 = 2 and there are no remainders. Students may use a different orientation to the example above.

Increasing to 11 counters, we can see that we have 1 remaining (the red counter below).

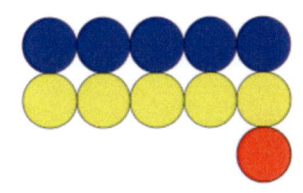

11 ÷ 5 = 2 r 1

Increasing to 12 counters, we can see that we have 2 remaining (the red counters below).

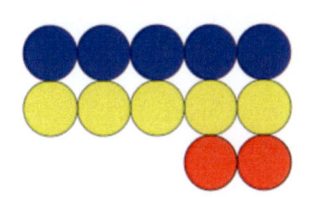

12 ÷ 5 = 2 r 2

Then with 13 and 14 counters, we can see we have 3 remaining and 4 remaining respectively.

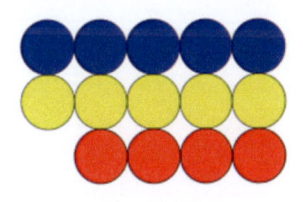

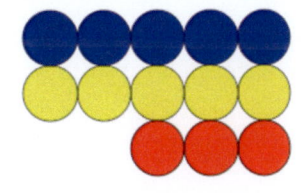

13 ÷ 5 = 2 r 3

14 ÷ 5 = 2 r 4

Now, when we increase to 15 counters, we can see that we have enough to make a new group of 5. The four remaining in the last example, has become 5 and it's a new group. There are no remainders.

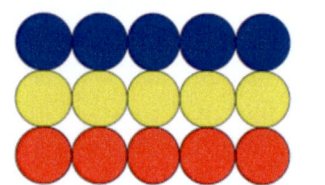

$15 \div 5 = 3$

Students should be encouraged to make conjectures based on the patterns they notice.

From the examples shown the statement is always true.

> **Notes:** Counters used in the examples are from MathsBot.com and the Number Lines are from Polypad.org.
>
> An alternative is to use number lines:
>
> Modelling $10 \div 5 = 2$ on a number line is shown here, clearly showing there is no remainder. 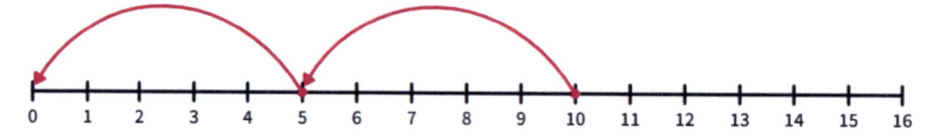
>
> The diagram below represents $11 \div 5 = 2 \, r \, 1$, the remainder identified as '1 left'. 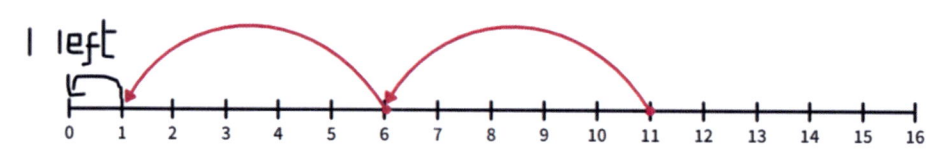
>
> The diagram below represents $12 \div 5 = 2 \, r \, 2$, the remainder identified as '2 left'. 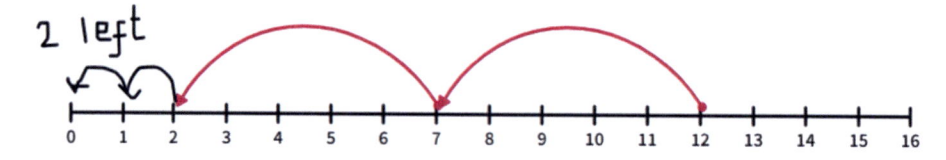
>
> The diagram below represents $13 \div 5 = 2 \, r \, 3$, the remainder identified as '3 left'.

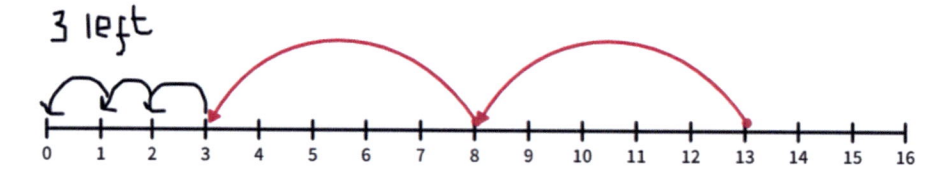

The diagram below represents 14 ÷ 5 = 2 r 4, the remainder identified as '4 left'.

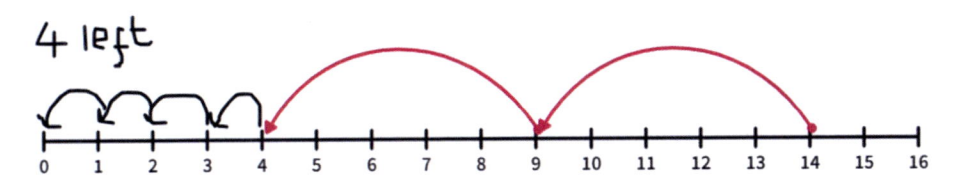

4 left

The final diagram showing no remainders when we reach the next multiple of 5, 15 ÷ 5 = 3.

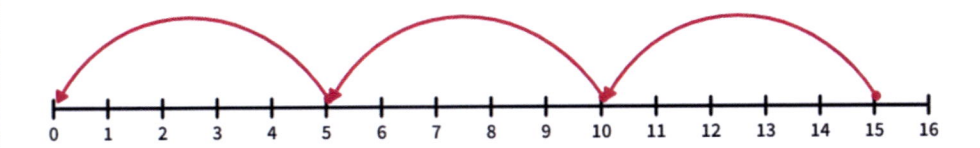

This is an opportunity to explore quotative (grouping or subtracting) and partitive division (sharing).

Students may further investigate what happens with other divisors. What is the largest remainder when dividing by 6 or 7 or 8.

MD016 Statement: An even number divided by an odd number gives a quotient with no remainder.

Answer: SOMETIMES TRUE

Manipulatives: Counters

This statement can be shown as sometimes true through using *Counters* and trying different numbers.

First of all, let's try some even numbers divided by 1, the first odd number. (Refer to MD014.)

6 ÷ 1 = 6 and there are no remainders.

8 ÷ 1 = 8 and there are no remainders.

10 ÷ 1 = 10 and there are no remainders.

These would all indicate the statement as being true.

Now let's try our divisor as 3:

6 ÷ 3 = 2 and there are no remainders.

8 ÷ 3 = 2 and there are 2 remaining.

10 ÷ 3 = 3 and there is 1 remaining.

These would lead us to see that the statement is not always true.

Students may use different orientations of the counters to represent the division. This is an opportunity to explore quotative (grouping or subtracting) and partitive division (sharing).

> **Notes:** Counters used in the examples are from MathsBot.com and the Number Lines are from Polypad.org.
>
> An alternative is to use number lines:
> Modelling 6 ÷ 1 = 6 on a number line is shown below, clearly showing there is one group (jump) and no remainder. 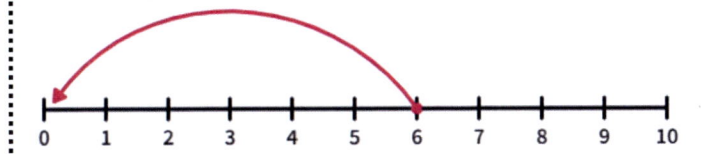
>
> The diagram below represents 8 ÷ 1 = 8. 
>
> Now with the same dividends but with a divisor of 3, the diagram below represents 6 ÷ 3 = 2 with two jumps and ending on zero. Therefore, there are no remainders. 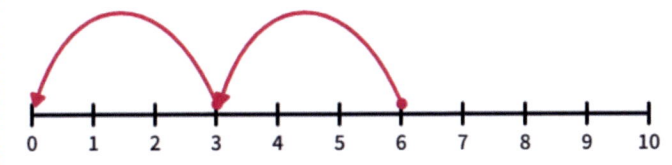
>
> The diagram below represents 8 ÷ 3 = 2 r 2. 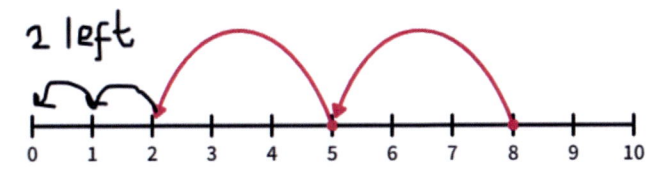
>
> Students may further investigate this statement: An odd number divided by an odd number gives a quotient with no remainder.

> **MD017 Statement:** When multiplying two whole numbers, the resulting product will be greater than either of the chosen numbers.
>
> **Answer:** SOMETIMES TRUE
>
> **Manipulatives:** Counters

This statement can be shown as sometimes true through using *Counters* and trying different examples.

Here we assume there is an understanding of the language and relationship:
product = factor x factor.

Let's first consider the number 10. The pairs of factors are: 1 and 10, 2 and 5. We can show these as rectangular arrays using counters.

Then we can say,

           10 = 1 x 10   and   10 = 2 x 5.

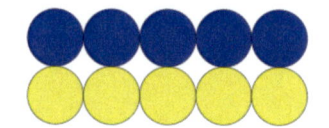

Clearly, 2 and 5 are both less than 10 indicating the statement might be true, but in the case of 1 and 10, while 1 is less than 10, 10 is equal to 10 and is therefore not greater.

Now consider another number, say 18. The pairs of factors are: 1 and 18, 2 and 9, 3 and 6.

18 x 1 = 18   (1 <18 however, 18=18)

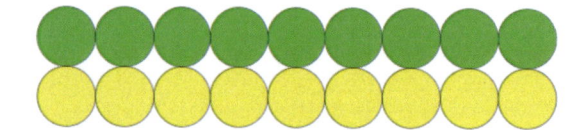

2 x 9 = 18   (2 < 18 and 9 < 18)

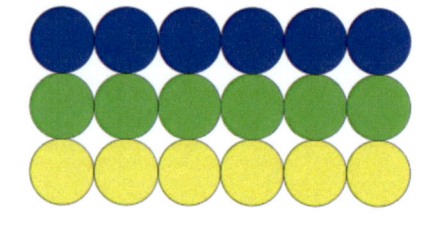

3 x 6 = 18 (3 < 18 and 6 < 18)

All these are less than 18 apart from 18 - the number itself.

Therefore, we would have to say that the statement is sometimes true.

Encourage students to be explicit in their conjectures. While there are many examples where the factors are both less than the product, there will always be an example where one of the factors will be equal to the product.

> **Notes:** Counters used in the examples are from MathsBot.com.
>
> This is an opportunity to explore commutativity and arrays.
>
> Other manipulatives such as maths link cubes or Cuisenaire rods could be used.

MD018 Statement: Halving a 3-digit number less than 200 gives a 2-digit answer.

Answer: SOMETIMES TRUE

Manipulatives: Place Value Counters, Dienes

Using Place Value Counters

This statement can be shown as sometimes true through using *Place Value Counters* by equally sharing the chosen number into two parts.

First of all, let's pick the largest three-digit number less than 200. This is 199.
Making this in counters can be shown as: (Students may use other ways of arranging the counters.)

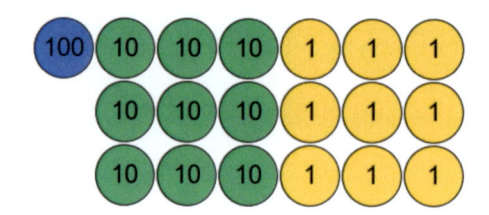

When dividing this into two equal parts, encourage the students to start from the largest place value. This supports the connection with the use of the division bracket.
The process that should be seen is described below:
Regrouping the 100 into ten of the 10s counters, and showing them shared into two equal groups, then sharing the others 10s counters. Thereafter, regrouping a 10s counter into ten of the 1s counters and then sharing out the remaining 1s counters:

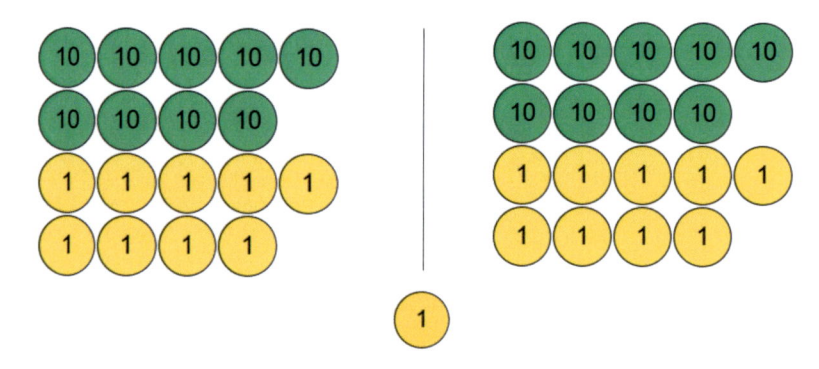

Here, the 1 remaining is split into two halves. Therefore, 199 ÷ 2 = 99 r 1 or 99 ½ or 99.5. So, this results in an answer with more than 2-digits.

Choosing the next number down would be 198. This divided by 2 is equal to 49 exactly, which is a 2-digit answer. (198 ÷ 2 = 99)

The above two examples show the statement is sometimes true.

Now consider the smallest 3-digit number: 100.

The 100 counter could be regrouped into ten 10s counters and divided by 2.

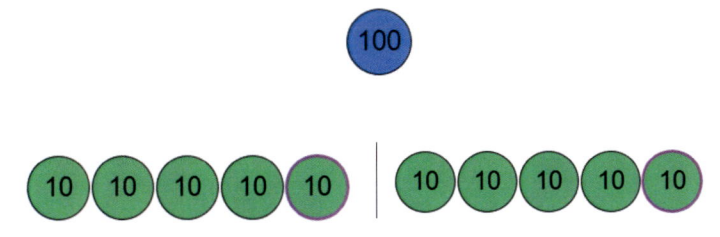

$100 \div 2 = 50$, which is a 2-digit number.

Numbers lower than this will be 2-digit numbers and not 3-digit numbers, therefore numbers smaller than a 100 are not considered.

Using Dienes

Dienes could also be used. Here the number 184 has been chosen.

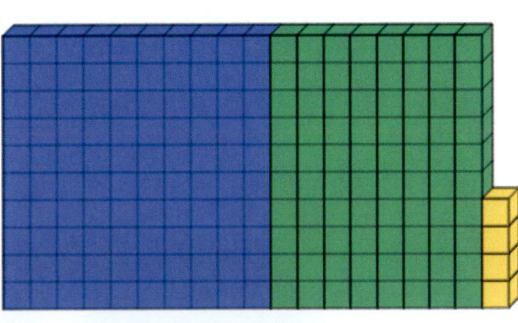

Then, partitioned using place value:

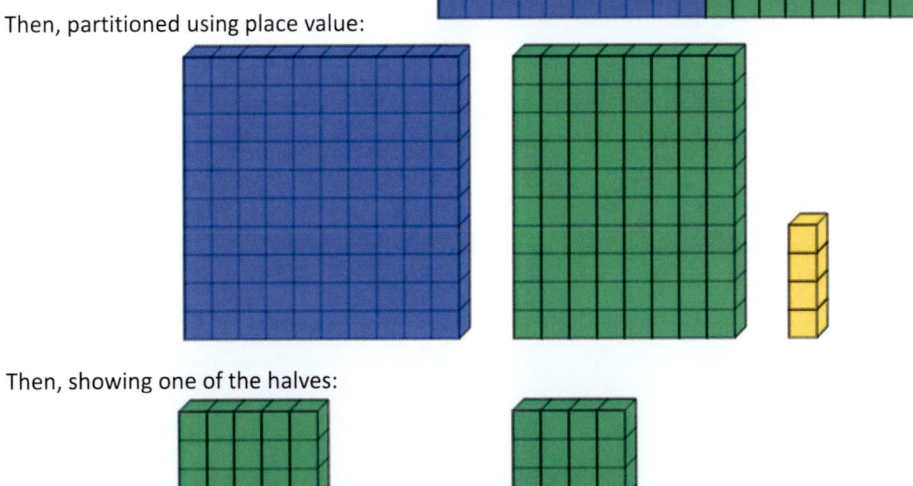

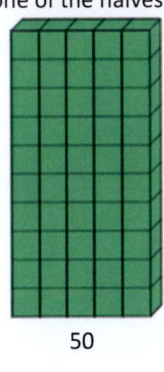

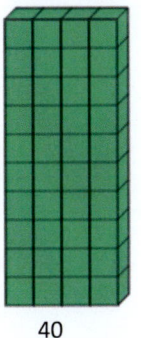

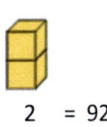

Then, showing one of the halves:

50 + 40 + 2 = 92

Encourage the students to show several examples of their own using their choice of manipulative.

> **Notes:** Place Value Counters, Hundred Square and Dienes used in the examples are from MathsBot.com.
>
> Using a 100 square with numbers beginning at 100 would also support children to see what the options were.
>
> | 100 | 101 | 102 | 103 | 104 | 105 | 106 | 107 | 108 | 109 |
> |---|---|---|---|---|---|---|---|---|---|
> | 110 | 111 | 112 | 113 | 114 | 115 | 116 | 117 | 118 | 119 |
> | 120 | 121 | 122 | 123 | 124 | 125 | 126 | 127 | 128 | 129 |
> | 130 | 131 | 132 | 133 | 134 | 135 | 136 | 137 | 138 | 139 |
> | 140 | 141 | 142 | 143 | 144 | 145 | 146 | 147 | 148 | 149 |
> | 150 | 151 | 152 | 153 | 154 | 155 | 156 | 157 | 158 | 159 |
> | 160 | 161 | 162 | 163 | 164 | 165 | 166 | 167 | 168 | 169 |
> | 170 | 171 | 172 | 173 | 174 | 175 | 176 | 177 | 178 | 179 |
> | 180 | 181 | 182 | 183 | 184 | 185 | 186 | 187 | 188 | 189 |
> | 190 | 191 | 192 | 193 | 194 | 195 | 196 | 197 | 198 | 199 |
>
> This is an opportunity to explore quotative (grouping or subtracting) and partitive division (sharing).
>
> Students may further investigate what happens when halving 3-digit numbers to 500 or more, or whatever other constraint they want to place.

MD019 Statement: Finding a third of a number is the same as dividing it by 3.

Answer: ALWAYS TRUE

Manipulatives: Counters with Bar Models

This statement can be shown as always true through using *Counters* and Bar Models and by considering times tables facts.

If we take 12 and divide this by 3, we get 4. Consider the 12 counters in the diagram here. They have been grouped into equal groups of three (shown by the rows). There are 4 rows so 12 ÷ 3 = 4 We can also know or say 3 x 4 = 12 and 4 x 3 = 12 and that 12 ÷ 4 = 3. The array reveals all of these.

To find a third of a number, we need to split the number into three equal parts. We can do this with a shape – as shown on this bar model.

For the number 12, we can also show it on a bar model like so. The whole is shown as 12. When this is divided into three equal parts – thirds, each third will be 4.

Trying another example, also a multiple of 3, we can show:
18 ÷ 3 = 6   (18 counters grouped into rows of 3 makes 6 rows.)

A third of 18 or $\frac{1}{3}$ of 18

Students could use counters and place all 18 on the whole bar, then share them equally between each of the thirds. They would end up with 6 counters placed on each third.

The examples shown reveal the statement to be always true for multiples of 3.

As not all numbers are multiples of 3, we suggest students explore the statement using other numbers too, which will have remainders. Encourage the students to explore how the remainder can also be divided by 3 or shared into 3 equal parts. This will reveal fractional or decimal answers.

Using consecutive numbers might be a helpful approach. For example:

$12 \div 3 = 4$ $\quad 13 \div 3 = 4\frac{1}{3}$ $\quad 14 \div 3 = 4\frac{2}{3}$ and demonstrate this with fraction bars or bar models.

> **Notes:** Counters, Hundred Square (below) and Bar Models used in the examples are from MathsBot.com.
>
> This is an opportunity to explore quotative (grouping or subtracting) and partitive division (sharing).
>
> Providing students with a grid such as that shown, allows students to make the connection between counting in 3's and the 3x multiplication table. It can be used alongside counters grouped into rows of 3.
>
> | 1 | 2 | 3 |
> |---|---|---|
> | 4 | 5 | 6 |
> | 7 | 8 | 9 |
> | 10 | 11 | 12 |
> | 13 | 14 | 15 |
> | 16 | 17 | 18 |
> | 19 | 20 | 21 |
> | 22 | 23 | 24 |
> | 25 | 26 | 27 |
> | 28 | 29 | 30 |
> | 31 | 32 | 33 |
> | 34 | 35 | 36 |
>
> Students may further investigate what happens whether finding a fifth is the same as dividing by 5, finding a sixth is the same as dividing by 6 and so on, grids showing the 5x and 6x table could support their thinking.

> **MD020 Statement:** Finding a quarter of a number is the same as halving and halving again.
>
> **Answer:** ALWAYS TRUE
>
> **Manipulatives:** Number Frames with Bar Models

This statement can be shown as always true through using *Number Frames* and *Bar Models* and trying different numbers divided by 2, and then divided by 2 again.

Beginning with 4 and using a number frame 4, then halving this and exchanging it for two of the twos, then halving again and getting 4 of the ones.

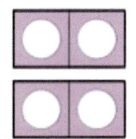

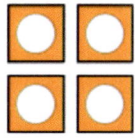

Hence, 4 ÷ 2 = 2 and then 2 ÷ 2 = 1

Students could check this with finding a quarter of 4 using bar modelling, and then seeing that this is 1.

Finding a quarter means dividing by 4:

Taking the whole, in this example 4, we can represent this on the top bar and then divide the bottom bar into 4 parts:
4 ÷ 4 = 1

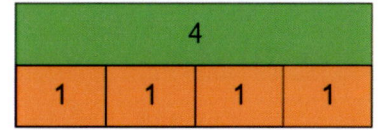

Students could also use bar modelling to show half and half again – like this:

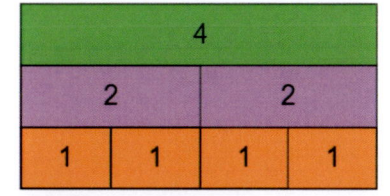

Showing this with another multiple of 4, e.g. 36.
We can make it with the number frames:
10, 10, 10 and 6. (36 = 10 + 10 + 10 + 6)

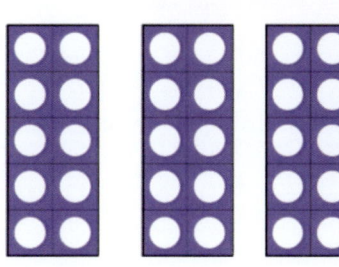

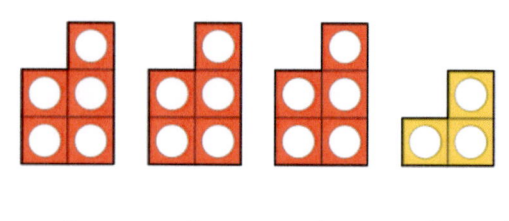

Halving each, we obtain:

$$5 + 5 + 5 + 3 = 18$$

We can regroup these for a 10 and an 8 as in the diagram here:

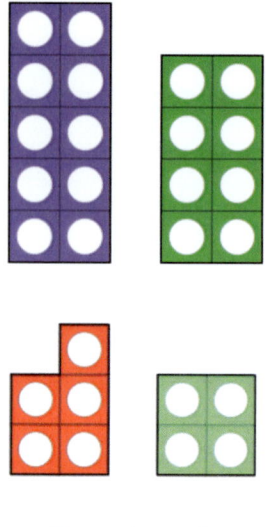

Now halving each of these we get:

$$5 + 4 = 9$$

Hence, 36 ÷ 2 = 18 and 18 ÷ 2 = 9.

Finding a $\frac{1}{4}$ of 36 can also be shown as:

Therefore, $\frac{1}{4}$ of 36 = 9, which is the same as 36 ÷ 4 = 9.

36			
9	9	9	9

We could try 42 as a further example. This is an even number but, it is not a multiple of 4.
42 ÷ 2 = 21 and 21 ÷ 2 = 10 remainder 1 or $10\frac{1}{2}$.
42 ÷ 4 = 10 with 2 remaining. If the two remaining are divided equally between 4 this gives $\frac{2}{4}$ or $\frac{1}{2}$.

Encourage students to consider the value of a remainder when the even number is not a multiple of 4.

Finally, students could investigate with some odd numbers. (It may be easier to use decimals at this point with the remainders, if fractional equivalents have not been discussed.)

For example, consider the number 23:
23 ÷ 2 = 11 remainder 1 or 11.5 and then 11.5 ÷ 2 = 5.75.

Using two 10 frames and a 3 frame to represent 23:

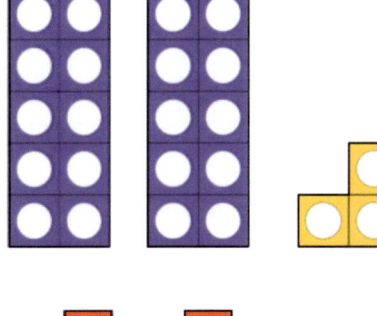

Halving the two 10 frames shows a 5 and a 5.

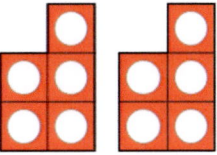

By regrouping the 3 frame for three 1 frames, we can see that half of 3 is equal to 1 and $\frac{1}{2}$:

From this we can see that 23 ÷ 2 = 11 $\frac{1}{2}$.

We can regroup the two 5 frames for a 10 frame. Using this to halve $11\frac{1}{2}$ can be seen in the diagram here.

Note that $\frac{1}{2}$ is shown by covering half of the one.

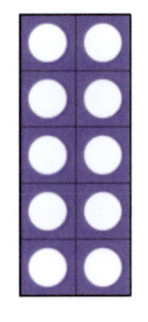

Halving these results in $11\frac{1}{2} \div 2 = 5$ and $\frac{3}{4}$, which is shown in the diagram on the right. Note that $\frac{1}{2}$ of 1 is shown by covering half of the one and $\frac{1}{2}$ of $\frac{1}{2}$ is shown by covering half of the half.

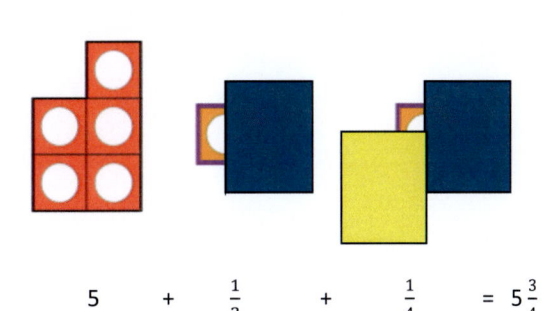

$$5 \quad + \quad \frac{1}{2} \quad + \quad \frac{1}{4} \quad = \quad 5\frac{3}{4}$$

The above examples show the statement to be always true.

> **Notes:** Number Frames, Hundred Square (below) and Bar Models used in the examples are from MathsBot.com.
>
> A 100-square with multiples of 4 marked could also be used.
>
> This is an opportunity to explore quotative (grouping or subtracting) and partitive division (sharing).
>
> This is also an opportunity to equate remainders with fractions and decimals in a controlled way, where there are either quarters, halves or three-quarters as part of the quotient.
>
> | 1 | 2 | 3 | 4 | 5 | 6 | 7 | 8 | 9 | 10 |
> |----|----|----|----|----|----|----|----|----|-----|
> | 11 | 12 | 13 | 14 | 15 | 16 | 17 | 18 | 19 | 20 |
> | 21 | 22 | 23 | 24 | 25 | 26 | 27 | 28 | 29 | 30 |
> | 31 | 32 | 33 | 34 | 35 | 36 | 37 | 38 | 39 | 40 |
> | 41 | 42 | 43 | 44 | 45 | 46 | 47 | 48 | 49 | 50 |
> | 51 | 52 | 53 | 54 | 55 | 56 | 57 | 58 | 59 | 60 |
> | 61 | 62 | 63 | 64 | 65 | 66 | 67 | 68 | 69 | 70 |
> | 71 | 72 | 73 | 74 | 75 | 76 | 77 | 78 | 79 | 80 |
> | 81 | 82 | 83 | 84 | 85 | 86 | 87 | 88 | 89 | 90 |
> | 91 | 92 | 93 | 94 | 95 | 96 | 97 | 98 | 99 | 100 |

MD021 Statement: The square of a number is greater than the number doubled.

Answer: SOMETIMES TRUE

Manipulatives: Cuisenaire Rods

This statement can be shown as sometimes true through using *Cuisenaire rods* and exploring with different numbers.

First of all, let's pick some numbers that show the statement to be true. Below are examples for the numbers 3 and 5 with numbers shown on the rods.

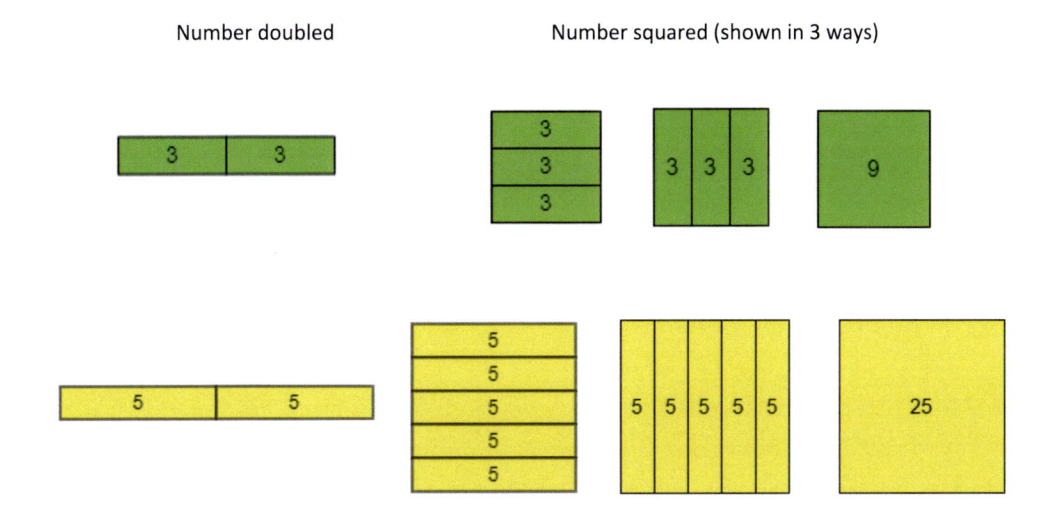

Number doubled   Number squared (shown in 3 ways)

For these two numbers, we can see that the square is bigger than a double, leading us to believe that the statement is true. We see that double 3 is 6 while squaring 3 is equal to 9, and double 5 is 10 while squaring 5 is 25.

Also shown here are the numbers 1 and 2 and how they might be modelled.

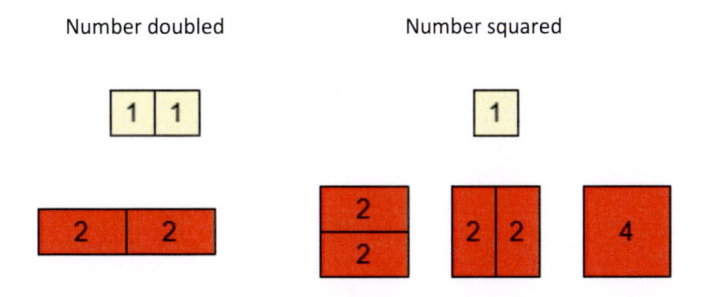

Number doubled   Number squared

What we notice here is that double 1 is 2 and this is bigger than the square of 1 which is 1. Also, that double 2 is 4 and this is the same as the square of 2 which is also 4.

Now we have examples where the statement is not true, so this leads us to conclude that the statement is only **sometimes** true.

> **Notes:** Cuisenaire Rods, Counters, Scaled times tables and Geoboard images used in the examples are from MathsBot.com.
>
> This is an opportunity to explore square numbers and how they do actually form a square shape with sides of equal length. Using counters – as illustrated below helps students to see the square outline that is formed.
>
> Choosing the number 3, we can take 3 counters and then double these to get 6 (3 + 3 or 3 x 2), while 3 squared is 9 (3 x 3). Shown in the diagrams here. 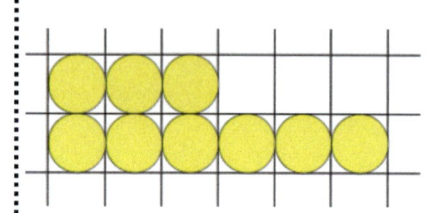
>
> Starting with '3' and looking at double '3' 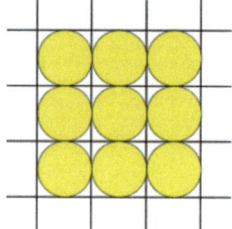   '3' squared – note the square shape
>
> Choosing numbers bigger than 3, also turns out to be true. Can students reason why?
>
> 4   4 x 2 = 8   while 4 x 4 = 16
> 5   5 x 2 = 10   while 5 x 5 = 25
> 6   6 x 2 = 12   while 6 x 6 = 36
> 12   12 x 2 = 24 while 12 x 12 = 144
>
> However, when using the number 1, 1 doubled is 2. 1 squared is 1 x 1 = 1. See diagram here: 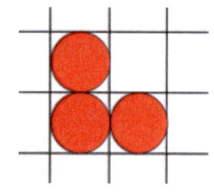 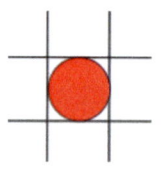
>
> So, the square of the number is smaller than double the number which leads to the statement above being not true. It is harder to see a single counter as a 'square shape' especially when the counter is circular. The focus should be on equal dimensions of width and length (array model). If available, square shaped counters would be better to show the 'squareness'.

Also, when considering the number 2, 2 doubled is 4. While 2 squared is 2 x 2 = 4. These are the same result.

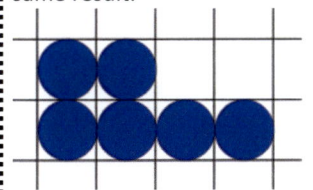

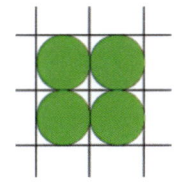

Taking '2' and doubling it

Taking '2' and squaring it

Therefore, the statement is sometimes true.

Considering scaled times tables – shows square numbers:

1	2	3	4	5
2	4	6	8	10
3	6	9	12	15
4	8	12	16	20
5	10	15	20	25

This is an opportunity to explore square numbers and how they do actually form a square shape with sides of equal length.

Using shape numbers:

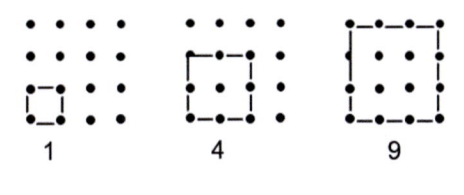

1   4   9

This is an opportunity to explore shape numbers and how they are formed through a two-dimensional shape with sides of equal length. For example, square numbers created using the shape as the structure as seen in the diagram here.

> **MD022 Statement:** A square number always has an odd number of factors.
> **Answer:** ALWAYS TRUE
> **Manipulatives:** Counters, Cuisenaire Rods

This statement can be shown as always true by using *Counters* and *Cuisenaire rods* and looking at factors through arrays.

Using Counters

First, let's look at the factors of 12 by arranging 12 counters into an array. The following arrays can be made.

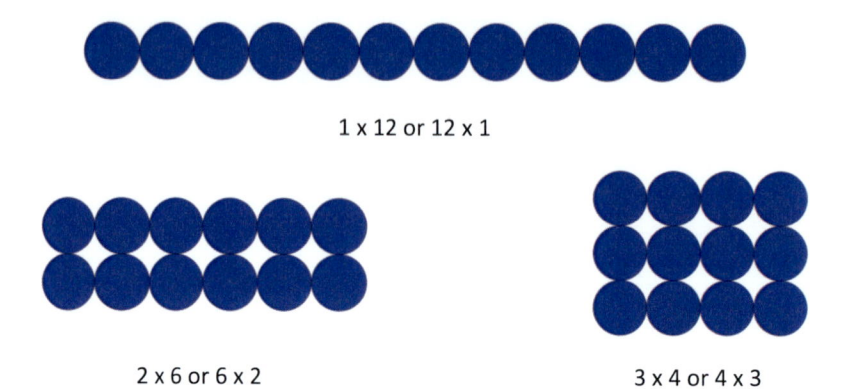

1 x 12 or 12 x 1

2 x 6 or 6 x 2   3 x 4 or 4 x 3

So, factors of 12 are: 1, 12, 2, 6, 3 and 4. and as they come in pairs, where the pairs are different to each other, there are therefore an even number of factors. Also, the arrays could be shown in other orientations, but the factor pairs remain the same.

Now for a square number, such as 9. The arrays can be modelled like this:

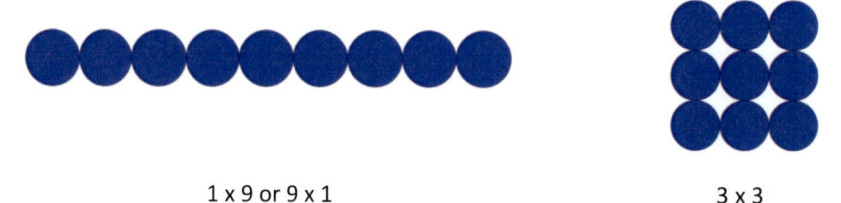

1 x 9 or 9 x 1   3 x 3

For the 3 x 3 array, square shaped counters would be better if available (refer to MD021). Now, because 3 x 3 makes a square number, both the factors are the same, in this case 3. Hence, the number of factors is odd. We have 1, 9 and 3 – just 3 factors.

Let's consider a few more square numbers:
1 x 1 = 1. So, just 1 factor (and an odd number).
2 x 2 = 4 and factors of 4 are 1, 4 and 2. So three factors here, again this is an odd number of factors.
4 x 4 = 16 and factors of 16 are 1, 16, 2, 8, 4. This has 5 factors, again an odd number.
5 x 5 = 25 and factors of 25 are 1, 25 and 5. This has 3 factors.

Using Cuisenaire Rods

Cuisenaire rods can also be used when students don't need to employ 'counting'.
So, on the pink rod we have the number sentence 1 x 4 = 4 and similarly with two red rods which can be moved to form a square: 2 x 2 = 4.

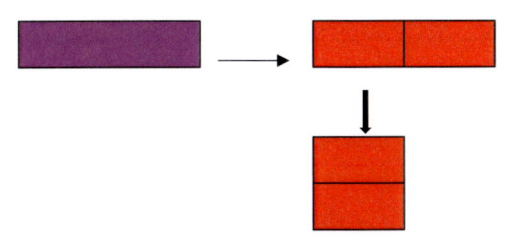

Hence the factors of 4 are: 1, 2 and 4.
This idea can be repeated in the same way for other square numbers.

From the examples shown the statement is always true.

> **Notes:** Counters and Cuisenaire Rods used in this example are taken from MathsBot.com.
>
> This is an opportunity to explore multiples and factors. A factor is a number that divides into another number a whole number of times, i.e. there is no remainder.
> For example, 4 is a factor of 12 as 12 ÷ 4 = 3.
> Another way to consider this is that factors of 12 can be listed as integers that multiply together to make exactly 12. Here, 3 x 4 = 12 so 3 and 4 are factors.
>
> Students can explore other types of numbers with their factors, for example: prime numbers will always have only 2 factors or the larger the number the greater the number of its factors.
>
> Another statement that could be investigated is "The larger the number the greater the number of its factors".
>
> First 20 Square numbers:
> 1, 4, 9, 16, 25, 36, 49, 64, 81, 100, 121, 144, 169, 196, 225, 256, 289, 324, 361, 400.

> **MD023 Statement:** Multiples of 6 are one more and one less than prime numbers.
>
> **Answer:** SOMETIMES TRUE
>
> **Manipulatives:** Counters, Hundred Square

This statement can be shown as sometimes true through using *Counters* and by considering times tables facts and primes on a *Hundred square grid.*

If we take 6 (shown in the middle of the diagram below) and then look at the numbers that are one less (5) and one more (7) we notice that these numbers are both prime. They only have the factors, one and themselves.

Considering 12, we notice again that the numbers either side of it, 11 and 13 are also prime. The number grid below shows multiples of 6 (in yellow) and primes (in red).

1	2	3	4	5	6	7	8	9	10
11	12	13	14	15	16	17	18	19	20
21	22	23	24	25	26	27	28	29	30
31	32	33	34	35	36	37	38	39	40
41	42	43	44	45	46	47	48	49	50
51	52	53	54	55	56	57	58	59	60
61	62	63	64	65	66	67	68	69	70
71	72	73	74	75	76	77	78	79	80
81	82	83	84	85	86	87	88	89	90
91	92	93	94	95	96	97	98	99	100

Looking at 18 (the next multiple of 6), we notice again that prime numbers sit on both sides. We can notice further examples too where the statement looks like it is true.

However, we can also see some cases where it is not true.

Looking at the number 24, we see that one less is a prime but one more is 25, whose factors are 1,5,25 and is therefore not a prime. The same is true for 36, one more is 37 (a prime) however, one less is 35, which has factors 1,5,7,35 and is therefore not prime.

Hence, the statement is sometimes true.

> **Notes:** Counters and Hundred Square used in the examples are from MathsBot.com.
>
> Using a six-column grid as shown below could be used as a scaffold to support students.
>
> | 1 | 2 | 3 | 4 | 5 | 6 |
> |----|----|----|----|----|----|
> | 7 | 8 | 9 | 10 | 11 | 12 |
> | 13 | 14 | 15 | 16 | 17 | 18 |
> | 19 | 20 | 21 | 22 | 23 | 24 |
> | 25 | 26 | 27 | 28 | 29 | 30 |
> | 31 | 32 | 33 | 34 | 35 | 36 |
> | 37 | 38 | 39 | 40 | 41 | 42 |
> | 43 | 44 | 45 | 46 | 47 | 48 |
> | 49 | 50 | 51 | 52 | 53 | 54 |
> | 55 | 56 | 57 | 58 | 59 | 60 |
>
> This is an opportunity to explore factors and prime numbers and students could also investigate the relationship between prime factors and the composition of the number.
> For example: 12 = 2 x 2 x 3
>
> Students may further investigate "All prime numbers (except 2 and 3) are within 1 of a multiple of 6" putting the emphasis on the prime numbers. This is much more challenging – and some notes follow here to support this.
>
> Let 6n represent a multiple of 6.
>
> The next numbers are: 6n+1, 6n+2, 6n +3, 6n +4, 6n +5, 6n + 6.
>
> Now, 6n and 6n + 6 are not prime as they are multiples of 6.
>
> 6n+2 is 2(3n +1) which is an even number and therefore not a prime.
> 6n + 4 is 2(3n+2) which is also even and not prime.
> 6n + 3 is 3(2n+1) which is a multiple of 3 and therefore not prime.
>
> So, this leaves 6n +1 and 6n +5:
> These are the only numbers that can be prime – they are not always prime – but therefore the statement is true – all prime numbers (except 2 and 3) are within 1 of a multiple of 6.

MD024 Statement: Dividing a number makes the quotient smaller than the dividend.

Answer: SOMETIMES TRUE

Manipulatives: Counters

This statement can be shown as sometimes true through using *Counters*.

First of all, let's remind ourselves of the vocabulary: dividend ÷ divisor = quotient.

So, for example, 14 ÷ 2 = 7 where 14 is the dividend, 2 the divisor and 7 the quotient. In this case, the quotient is smaller than the dividend, as 7 is less than 14.

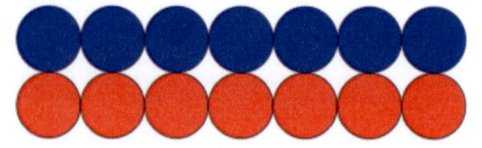

The array above might represent 14 divided into two groups – shown by the two colours.

Further examples also support this.

6 ÷ 3 = 2 and 2 is less than 6    8 ÷ 4 = 2 and 2 is less than 8

6 divided into 3 groups (shown with 3 colours) and 8 divided into 4 groups (shown with 4 colours). In each of the cases above, the divisor is less than the dividend.

Now let's consider what happens when the divisor is the same as the dividend. Does this change the result?

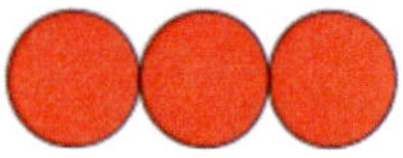

3 ÷ 3 = 1 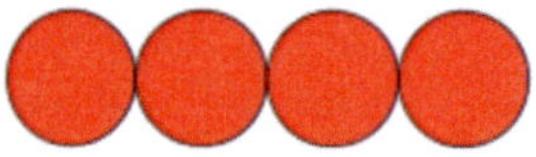   4 ÷ 4 = 1 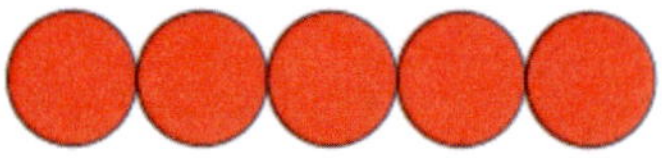   5 ÷ 5 = 1

These examples show the quotient is 1, and this is smaller than the dividend. So, the statement still seems to be true.

This might be where students conclude that the statement is always true and that might be fine depending on the age and the understanding of the students concerned. Students should be encouraged to state under what conditions it is always true. This will support them looking at different types of numbers. For example, what happens when the divisor is 1 or the divisor is larger than the dividend, or the divisor is a fraction?

Now let's consider what happens when the divisor is 1. The diagrams above can help to illustrate these too.

$$3 \div 1 = 3 \qquad 4 \div 1 = 4 \qquad 5 \div 1 = 5$$

In these cases, the quotient is the same as the dividend, it is neither bigger or smaller. This leads us to see that the statement is not always true.

What happens when we consider a divisor that is bigger than the dividend? This will lead to quotients which are not integers.

Using the example 3 ÷ 6 or "Can I share 3 into 6 equal parts?" Some students will relate this division to a fraction and will recognise it as $\frac{3}{6}$ or $\frac{1}{2}$. This can be seen by using 3 counters and imagining halving each of them to create 6 equal size pieces as shown in the diagram here.

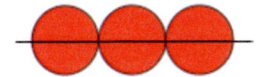

$3 \div 6 = \frac{3}{6} = \frac{1}{2}$ The quotient is smaller than the dividend.

Taking the example of 3 ÷ 4, using an example familiar to the students such as pizza or cake can help them to visualise 3 shared between 4. How much will each person get?

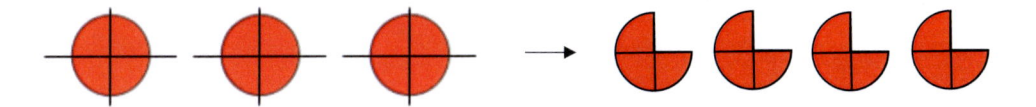

$3 \div 4 = \frac{3}{4}$ The quotient is smaller than the dividend.

However, there may be some who will consider what happens when the divisor is a fraction.

Consider $8 \div \frac{1}{2}$: This could be read as how many halves are there in 8?

With the help of 8 counters, we can see this is 16. So, $8 \div \frac{1}{2} = 16$ and the quotient is bigger than the dividend.

Another example shown here is $3 \div \frac{1}{4}$ is equal to:

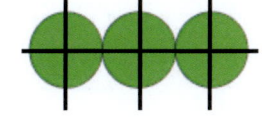

The number of quarters in 3 is 12. $3 \div \frac{1}{4} = 12$ and again the quotient is bigger than the dividend.

This all leads to the conclusion that the statement is sometimes true.

Notes: Counters in the examples are from MathsBot.com.

Other manipulatives such as Dienes or maths link cubes maybe used as an alternative to counters. Cuisenaire rods provide an opportunity to avoid recounting one by one where that scaffold is not needed.

This is an opportunity to explore quotative (grouping or subtracting) and partitive division (sharing) and the language around division. In particular, when dividing by fractions, the focus on the language will support the understanding of the concept of dividing by a number smaller than 1.

Students may further investigate "Dividing a number makes it bigger" or "Taking two different numbers, say m and n, the quotient of m and n will equal the quotient of n and m (m ÷ n = n ÷ m)." This will allow students to explore whether division is commutative or not.

> **MD025 Statement:** Any whole number can be multiplied by partitioning (e.g. 35 x 4 is 30 x 4 plus 5 x 4).
>
> **Answer:** ALWAYS TRUE
>
> **Manipulatives:** Dienes, Place Value Counters

This statement can be shown as always true through using *Dienes* and *Place Value Counters* and is in fact what is known as the "distributive law".

Using Dienes

Let's look at 35 x 4 by partitioning the 35 into 30 and 5 and multiplying each of these by 4.

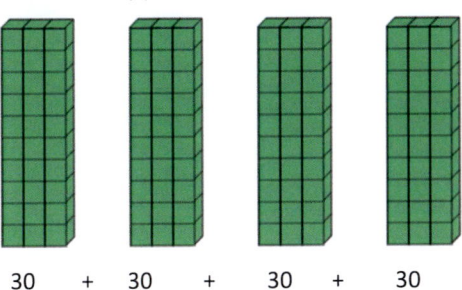

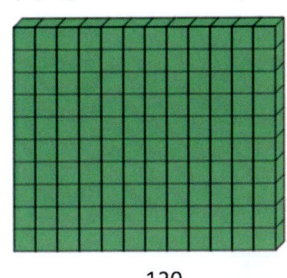

30 + 30 + 30 + 30 = 120

This can be regrouped into:

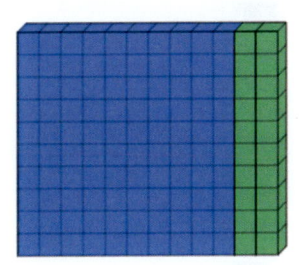

So, 30 x 4 = 120 as shown in the diagram above.

Now consider 5 x 4:

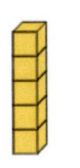

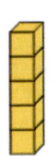

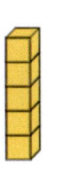

5 + 5 + 5 + 5 = 20 or 20

Summing the two products, 35 x 4 = 30 x 4 + 5 x 4
= 120 + 20
= 140.

Students should try several examples to see if this always works.

Using Place Value Counters

Using Place Value Counters and an array structure also helps to show visually how multiplication distributes over addition.

The example 23 x 5 is shown as 23 multiplied 5 times - where 23 can be seen repeated in each of the 5 rows.

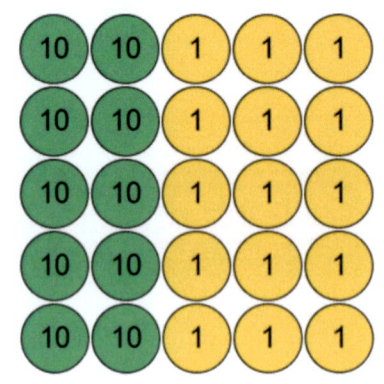

Students might consider how this multiplication is written using repeated addition and how it can be rewritten using partitioning and the rule of associativity.

23 x 5 = 23 + 23 + 23 + 23 + 23

= 20 + 3 + 20 + 3 + 20 + 3 + 20 + 3 + 20 + 3

= 20 + 20 + 20 + 20 + 20 + 3 + 3 + 3 + 3 + 3

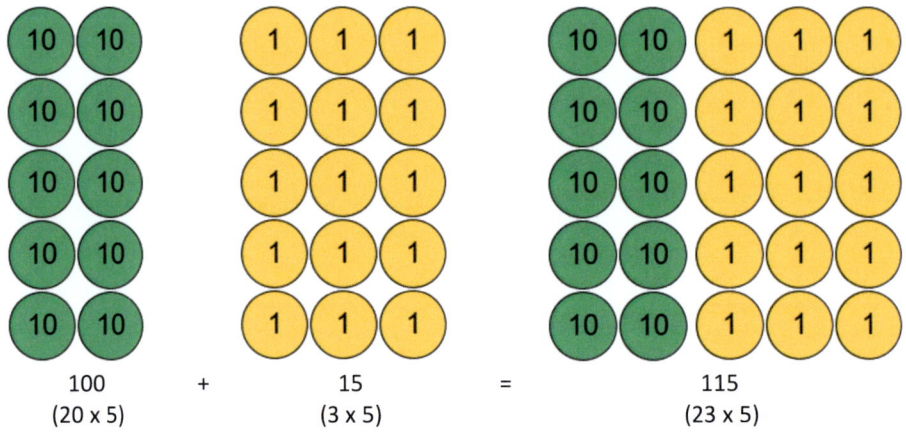

= (20 x 5) + (3 x 5)

100   +   15   =   115
(20 x 5)   (3 x 5)   (23 x 5)

Students could also explore other ways to partition the two-digit number, 23.

For example: 23 = 12 + 11, therefore 23 x 5 = (12 x 5) + (11 x 5).

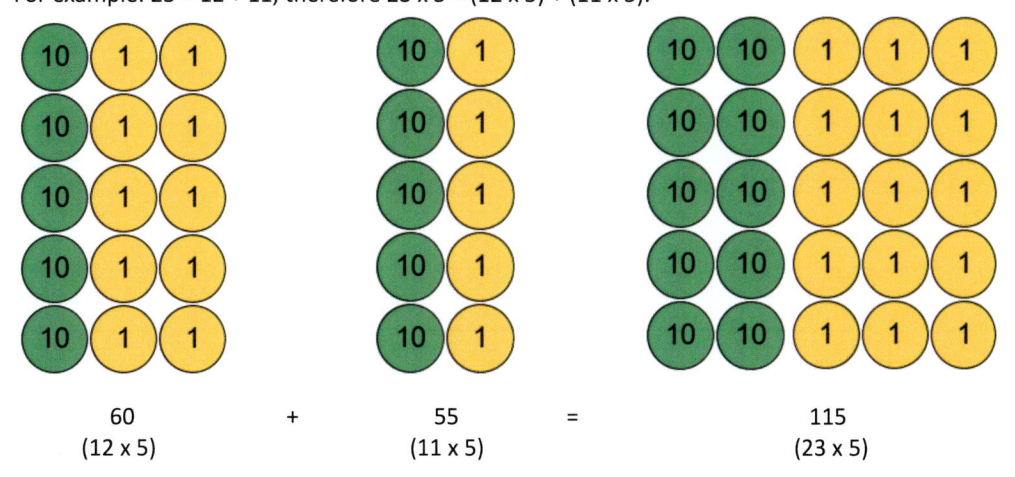

| 60 | + | 55 | = | 115 |
| (12 x 5) | | (11 x 5) | | (23 x 5) |

From the examples shown the statement is always true.

> **Notes:** Dienes and Place Value Counters used in the examples are from MathsBot.com.
>
> This is an opportunity to explore the rules of associativity, commutativity and distributivity.
>
> Commutative law: The numbers can be swapped over, and the result is the same e.g.:
> 5 + 2 = 2 + 5 = 7 and 5 x 2 = 2 x 5 = 10. This works for addition and multiplication and **not** for subtraction or division.
>
> Associative rule: The numbers can be grouped differently. It doesn't matter what is calculated first e.g. (2 + 3) + 4 = 2 + (3 + 4) = 9 and (2 x 3) x 4 = 2 x (3 x 4) = 24. This works for addition and multiplication and **not** for subtraction or division.
>
> Distributive rule: 3 lots of (2 + 4) is the same as 3 lots of 2 plus 3 lots of 4 or written mathematically, 3 x (2 + 4) = (3 x 2) + (3 x 4). This means 3 x can be "distributed" across the 2 + 4 into 3 x 2 and 3 x 4. The multiplication is distributed over the addition.
>
> Another example of partitioning a single digit number with counters is shown here, where 9 x 4 = (6 x 4) + (3 x 4). 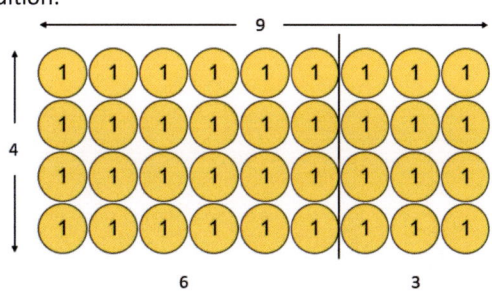
>
> From a general point of view:
> 4(a + b) = 4a + 4b and 8(a + 2b) = 8a + 16b
> Similarly, 3(a − b) = 3a − 3b, so, this works for addition and subtraction however does **not** work for division.
>
> Students may wish to explore why the distributive law does not work with division.
>
> Students may explore the validity of the statement for other types of numbers.

Chapter 3

Fractions, Decimals and Percentages Statements

The larger the denominator of a fraction, the larger the fraction. FD001	If the numerator and the denominator are the same, then it is equal to one whole. FD007
When the denominator is twice the numerator the fraction is worth 0.5. FD002	Fractions are less than one whole. FD008
The numerator of a fraction is less than the denominator. FD003	Doubling both numerator and denominator creates a new fraction that is double in size to the original one. FD009
Finding a fifth of a quantity is the same as multiplying by ten and halving the result. FD004	Dividing a fraction by 2 is the same as multiplying it by $\frac{1}{2}$. FD010
The sum of two or more fractions can be calculated by adding the numerators and the denominators independently. FD005	10% of a number is the same as the number multiplied by 0.1. FD011
Fractions with the same denominator can be ordered in size. FD006	Finding 5% of a number is the same as dividing the number by 5. FD012

30% of 80 is the same as 80% of 30. FD013	When comparing decimals, the longer the decimal the bigger the number. FD016
Reducing a number by 10% followed by increasing it by 10% results in returning to the original number. FD014	Decimal numbers can be written as fractions. FD017
A quarter ($\frac{1}{4}$) can be written as 1.4. FD015	Fractions can be simplified. FD018

> **FDP001 Statement:** The larger the denominator of a fraction, the larger the fraction.
>
> **Answer:** SOMETIMES TRUE
>
> **Manipulatives:** Fraction Wall

This statement can be shown as sometimes true through the use of a *Fraction wall* by comparing fractions.

First let's consider unit fractions:
In the diagram below, the lengths vividly show the statement to be not true. Notice that the assumption here is that 1 whole can be interpreted as $\frac{1}{1}$.

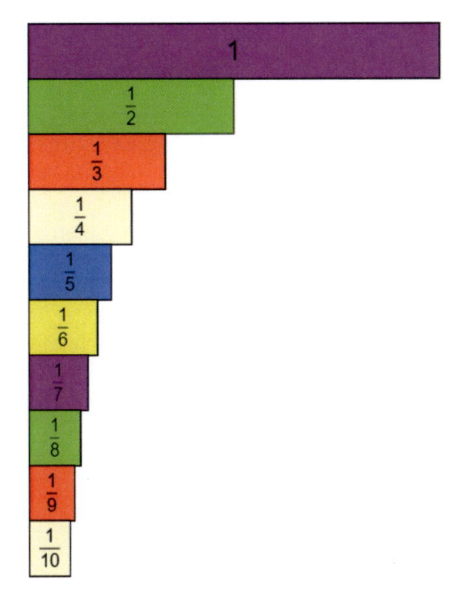

Students may express the concrete in abstract as $1 > \frac{1}{2} > \frac{1}{3} > \frac{1}{4} > \frac{1}{5} > \frac{1}{6} > \frac{1}{7} > \frac{1}{8} > \frac{1}{9} > \frac{1}{10}$.
This might lead students to predict that the statement is never true.

Now, let's consider non-unit fractions:
To explain this thinking consider comparing $\frac{1}{2}$ with increase of unit fractions of $\frac{1}{5}$s. See the diagrams below.

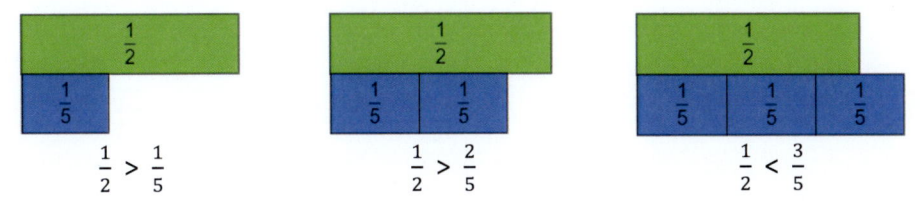

$\frac{1}{2} > \frac{1}{5}$ $\frac{1}{2} > \frac{2}{5}$ $\frac{1}{2} < \frac{3}{5}$

For the example above, it can be seen visually that the statement is sometimes true with the physical manipulation of the fractions. However, as mathematicians our aim is to understand why this is the case and even more so, to understanding the conditions under which the statement is sometimes true.

An alternative approach is to employ the idea of equivalent fractions. Using the example as above, it can be visualised as seen in the diagrams below. Therefore, the connection with equivalence and comparisons provides the evidence to support the given cases.

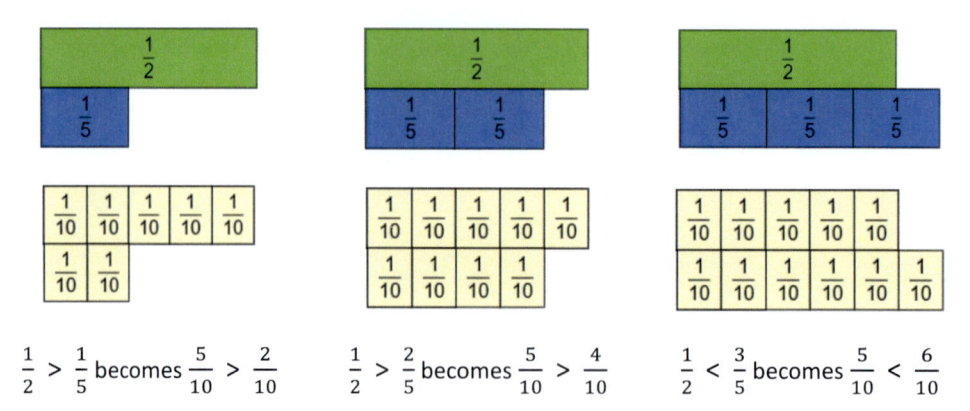

$\frac{1}{2} > \frac{1}{5}$ becomes $\frac{5}{10} > \frac{2}{10}$ $\frac{1}{2} > \frac{2}{5}$ becomes $\frac{5}{10} > \frac{4}{10}$ $\frac{1}{2} < \frac{3}{5}$ becomes $\frac{5}{10} < \frac{6}{10}$

This reveals that the statement could be true. Hence, considering all of the above, the statement is sometimes true.

Encourage students to explore with many examples to reach their own conclusions including examples where the numerator is the same value, but the denominator varies; for example, compare $\frac{4}{7}$ and $\frac{4}{5}$.

Notes: Fraction Wall in the examples are from MathsBot.com.

This statement could be investigated using different types of fractions besides unit fractions and proper fractions, where the principal idea of understanding the size of a fraction using a common denominator remains the same.

The first diagram which represents the statement: $1 > \frac{1}{2} > \frac{1}{3} > \frac{1}{4} > \frac{1}{5}$ ... can be explored further by identifying a common denominator to compare equivalent fractions. Note that the least common multiple for 1,2,3,4,5,6,7,8,9,10 is 2520. As $\frac{1}{2520}$ will be tricky to visualise, students can be shown 1mm as $\frac{1}{1000}$ on a metre stick to get a sense of how small $\frac{1}{2520}$ could be.

FDP002 Statement: When the denominator is twice the numerator, the fraction is worth 0.5

Answer: ALWAYS TRUE

Manipulatives: Fraction Wall

This statement can be shown as always true through the use of *Fraction walls*. To investigate this statement students should be familiar with the vocabulary associated with fractions and able to show their understanding of $\frac{1}{2}$ and its connection to its decimal equivalent value of 0.5.

The approach shown here is through the idea of equivalence. In the diagrams below, the whole has been shown as a comparison in relation to a half.

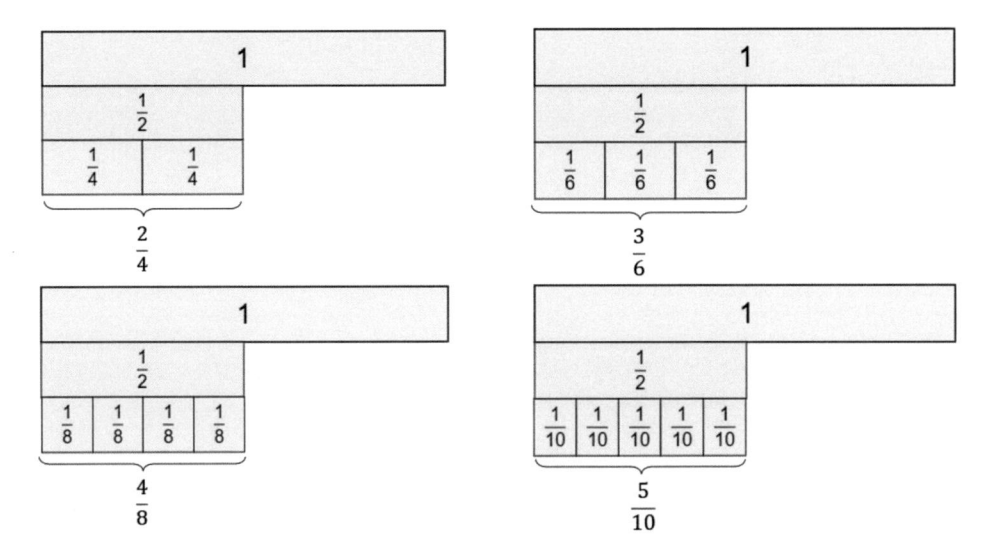

Encourage students to make jottings that show the equivalence relationship.
For example: $\frac{2}{4} = \frac{1}{2} = 0.5$, $\frac{3}{6} = \frac{1}{2} = 0.5$, $\frac{4}{8} = \frac{1}{2} = 0.5$, $\frac{5}{10} = \frac{1}{2} = 0.5$.

Notes: Fraction Walls in the examples are from MathsBot.com.

Further discussions of equivalence between non-unit fractions can support the connection with factors and multiples.

The visualisation of the sizes of the bars in the fraction wall is an important element in this activity. Students may use vocabulary such as double and twice interchangeably as well as half and 'the same'. Encourage the precise use of mathematical language alongside the concrete representation of the fractions.

Where a fraction wall is not accessible, it can be created using strips of paper of equal height and width, where the values assigned are in relation to the folds or to the equal parts per strip.

> **FDP003 Statement:** The numerator of a fraction is less than the denominator.
>
> **Answer:** SOMETIMES TRUE
>
> **Manipulatives:** Fraction Wall

This statement can be shown as sometimes true through the use of a *Fraction wall*.

In all the examples below, we have considered using strips of paper as the fraction wall. Therefore, for ease of folding, an even number of equal parts in relation to one whole is used. That is not to say that this investigation cannot be carried out for an odd number of equal parts. Note that students should be familiar with the terminology associated with fractions, such as unit fractions, proper fractions, whole numbers, improper fractions, numerator, denominator, vinculum (the horizontal line between the numerator and denominator).

First, let us consider unit fractions – When the fractional value is written to represent its size as in the diagram below, the statement appears to be true. The numerators are all smaller than their denominator parts. Notice that, here the one whole has been placed for the purpose of identifying the fractional parts in relation to that one whole.

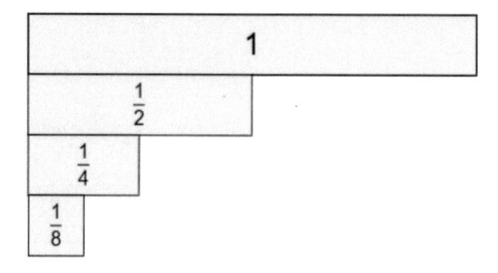

At this stage students may predict that the statement is true for all fractions. Encourage the students to investigate other fractions which are not unit fractions.

Now, consider fractions where the value of the numerator is equal to the value of the denominator. In the diagram below, the jottings can be written next to the bars to support the findings. Discuss the connection between such fractions and one whole. The students may connect the idea of repeated addition to multiplication in relation to fractions.

$$\frac{1}{2}+\frac{1}{2}=\frac{2}{2}$$

$$\frac{1}{4}+\frac{1}{4}+\frac{1}{4}+\frac{1}{4}=\frac{4}{4}$$

$$\frac{1}{8}+\frac{1}{8}+\frac{1}{8}+\frac{1}{8}+\frac{1}{8}+\frac{1}{8}+\frac{1}{8}+\frac{1}{8}=\frac{8}{8}$$

For this condition, the statement appears to be not true. This shows that fractions can have numerators equal in value to their denominators.

Now, consider improper fractions where the value of the numerator is greater than the value of the denominator. Here the approach is by increasing the size of the fractions above by an additional fractional part respectively as seen in the diagram below. Again, the jottings can be written next to the bars to support the findings. Discussion of the connection between such fractions and a whole leads nicely to the relationship between improper fractions and mixed fractions.

$$\frac{1}{2}+\frac{1}{2}+\frac{1}{2}=\frac{3}{2}$$

$$\frac{1}{4}+\frac{1}{4}+\frac{1}{4}+\frac{1}{4}+\frac{1}{4}=\frac{5}{4}$$

$$\frac{1}{8}+\frac{1}{8}+\frac{1}{8}+\frac{1}{8}+\frac{1}{8}+\frac{1}{8}+\frac{1}{8}+\frac{1}{8}+\frac{1}{8}=\frac{9}{8}$$

For this condition, the statement appears to be not true, as well.

From all the examples shown the statement is sometimes true.

Notes: Fraction Wall in the examples are from MathsBot.com.

Where a fraction wall is not accessible, it can be created using strips of paper of equal height and width, where the values assigned are in relation to the folds or to the equal parts per strip.

A unit fraction is a fraction where the numerator is 1 and the denominator is greater than 1.

A proper fraction is a fraction whose numerator cannot be greater nor equal to the denominator.

One whole is a fraction where both the numerator and denominator are equal.

An improper fraction is a fraction whose numerator is greater than its denominator.

A mixed fraction is a fraction with a whole number and a proper fraction.

> **FDP004 Statement:** Finding a fifth of a quantity is the same as multiplying by ten and halving the result.
>
> **Answer:** NEVER TRUE
>
> **Manipulatives:** Cuisenaire Rods

This statement can be shown as never true through the use of *Cuisenaire rods* by assigning a quantity to a given bar and manipulating it using the knowledge of part whole. This can be represented as shown in the diagram below:

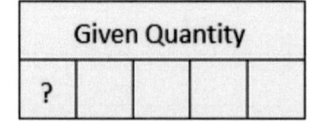

For example: If the given quantity is 40g then, the information can be represented as seen in the diagram on the right. Note that the '? – question mark' relates to a fifth of the given quantity and that it is not essential to be marked on the extreme left part as shown, it can be marked on any of the other equal parts. Here the equality in the size of the part is essential for students to understand. (See diagrams in notes.)

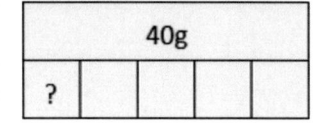

Another representation of the transition from the above model of the Cuisenaire rods could be through the area model:

Both representations should reveal the calculation: 40g ÷ 5 = 8g.

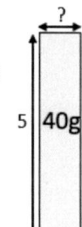

The statement suggests that the calculation should be to multiply the given quantity by 10, then half the result. Applying this to the example provided can be represented using the area model with the Cuisenaire rods (see diagram below).

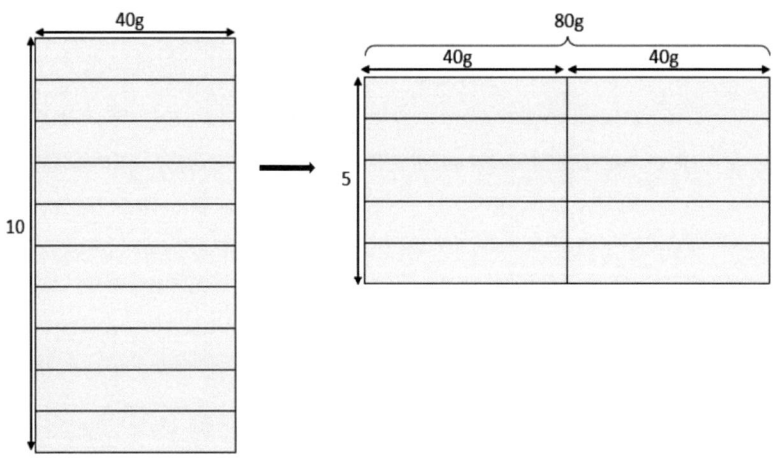

The arrow in the diagram shows the transition from multiplying by 10 then halving:
40g x 10 ÷ 2 = 400g ÷ 2 = 200g.

However, the final part of the diagram places the focus on the area revealing the calculation to be 400g ÷ 5 = 80g.

As 80g ≠ 200g, this shows that the statement is not true.

Students should be encouraged to investigate this statement using different quantities.

Students can continue to investigate by conjecturing alternative methods for dividing by 5. This will support making connections with multiplying and dividing fractions as well as with equivalent fractions.

For example: $\frac{1}{5} \times 40g = 8g$ is the same as $\frac{1}{5} \times \frac{40}{1} = \frac{1 \times 40}{5 \times 1} = \frac{40}{5} = 8$.

Using the knowledge of equivalence:

$\frac{1}{5} = \frac{2}{10}$, $\frac{2}{10} \times 40g = 8g$ is the same as $\frac{2}{10} \times \frac{40}{1} = \frac{2 \times 40}{10 \times 1} = \frac{80}{10} = 8$.

This is translated into a fifth of a given quantity can be calculated by multiplying by 2 and dividing the result by 10 or a fifth of a given quantity can be calculated by dividing by 10 and multiplying the result by 2.

Notes: Cuisenaire Rods in the examples are from MathsBot.com.

For the example above, the representation of which part is equal to a fifth may be seen like this:

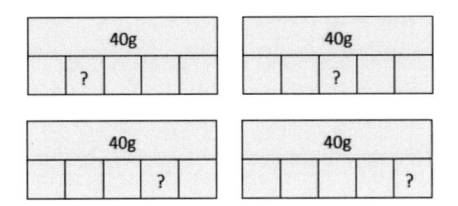

The visualisation with the rods is an important element in this activity, particularly in transition from multiplying by 10 and then dividing by 2. Notice the constant value of the area supports conceptualising the operations at play.

Students may further investigate other such relationships with unit fractions and their equivalent fractions. For example: a quarter of a given quantity is the same as multiplying by 2 then dividing the result by 8 or dividing by 8 and multiplying the result by 2.

> **FDP005 Statement:** The sum of two or more fractions can be calculated by adding the numerators and denominators independently (e.g.: $\frac{1}{2} + \frac{1}{3} = \frac{1+1}{2+3}$).
>
> **Answer:** NEVER TRUE
>
> **Manipulatives:** Fraction Wall

This statement can be shown as never true through using a *Fraction wall* such that there is a common understanding that each fraction is related to a whole by being a part of that whole.

Note that the act of physically joining the individual fractional parts is assumed to result in the sum of those parts, as shown in the diagrams below. Students should be encouraged to provide several names given to the numerical value of the sum. This will accommodate for the connections within fractions. For example, another name for two halves is one whole; another name for three halves is one whole and a half etc.

First let's consider fractions with the same denominator:

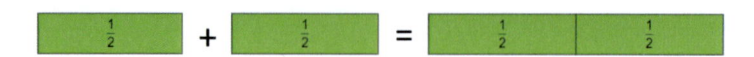

One half plus one half is equal to two halves, $\frac{1}{2} + \frac{1}{2} = \frac{2}{2}$. Here it can be seen that $\frac{1}{2} + \frac{1}{2} \neq \frac{2}{4}$. Furthermore, students may identify that $\frac{2}{4} = \frac{1}{2}$, showing that the sum cannot be a half.

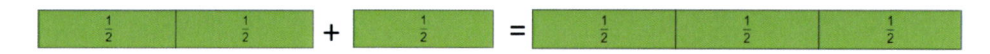

Two halves plus one half is equal to three halves, $\frac{2}{2} + \frac{1}{2} = \frac{3}{2}$. Here it can be seen that $\frac{2}{2} + \frac{1}{2} \neq \frac{3}{4}$. Furthermore, students may compare three quarters to three halves physically to show that they are not the same in size.

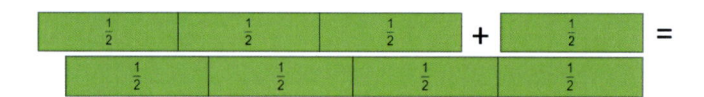

Three halves plus one half is equal to four halves, $\frac{3}{2} + \frac{1}{2} = \frac{4}{2}$. Here it can be seen that $\frac{3}{2} + \frac{1}{2} \neq \frac{4}{4}$. Furthermore, students may identify that four quarters is the same as one whole, thus showing that the sum cannot be one whole.

Now, consider fractions with different denominators:

As the denominators are different, students should be encouraged to understand that equivalent fractions with the same denominator of the fractions in the question are used to enable the addition.

The diagram on the following page shows the use of the lowest common multiple of both denominators.

One half plus one third is equal to three sixths plus two sixths.

Therefore, one half plus one third is equal to five sixths, $\frac{1}{2}+\frac{1}{3}=\frac{3}{6}+\frac{2}{6}=\frac{5}{6}$. Here it can be seen that $\frac{1}{2}+\frac{1}{3} \neq \frac{2}{5}$. Furthermore, students may compare two fifths to five sixths physically to show that they are not the same in size.

Finally, let's consider fractions with the numerators and denominators with equal value:
For example: $\frac{1}{1}+\frac{1}{1}=\frac{1+1}{1}=\frac{2}{1}=2$.
However, the statement suggests that $\frac{1}{1}+\frac{1}{1}=\frac{1+1}{1+1}=\frac{2}{2}=1$, which is not true.

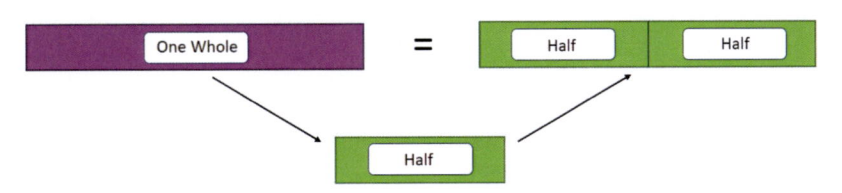

> **Notes:** Fraction Wall used in the examples are from MathsBot.com.
>
> If strips of paper (of equal height and width) are used, then the assumption here is that the student has a good understanding of constructing a fraction in relation to a whole as a concrete model. For example, the diagram below shows the manipulation with a strip of paper to show how a half can be created and connected to one whole. Let a strip of paper be called one whole (purple), then fold it into two equal parts (arrow from purple to green) then, open the fold to reveal two halves which are equal to one whole.
>
> Appreciate that there is a skill involved in folding strips of paper into equal parts. However, when comparing fractions with different denominators ensure the whole used is kept the same. Therefore, in the case of using strips of paper, ensure that the same size strip of paper is used.
>
> Further investigations with whole numbers and fractions can be carried out in a similar way as whole numbers can be expressed as a fraction.

> **FDP006 Statement:** Fractions with the same denominator can be ordered in size.
>
> **Answer:** ALWAYS TRUE
>
> **Manipulatives:** Fraction Wall and Number Line

This statement can be shown as true through the use of a *Fraction wall* such that there is a common understanding that each fraction is related to a whole by being a part of that same whole.

Assuming that the fractions are created using multiples of unit fractions, then the diagrams below show their size ordered in increments of unit fractional amount. Notice that the fraction wall shows the area of the bar whereas the number line shows the location of the fraction.

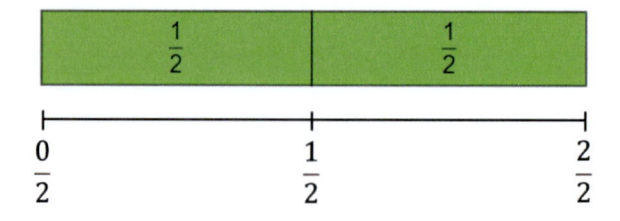

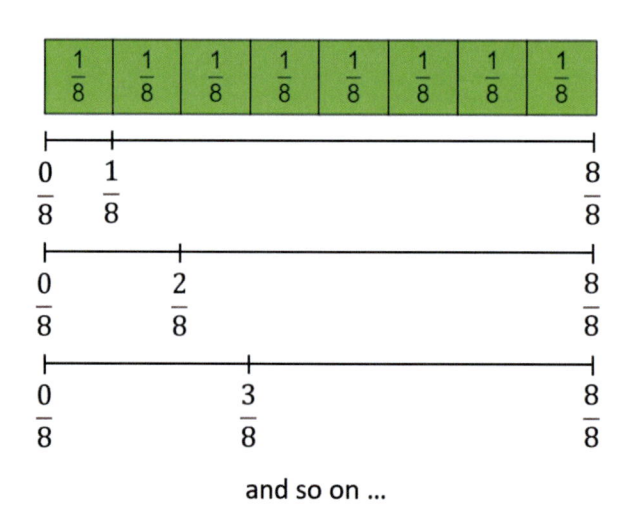

and so on ...

Students may express their understanding of the size of whole numbers (especially in consecutive order) to support their ordering of fractions. In this case, emphasizing that whole numbers can be written as fractions can support recognising a pattern.

For example, a number line that represents whole numbers can be written in fractional form as in the diagram here:

Where students write the number lines with unit fractions there is an opportunity to discuss equivalent fractions and how they can be represented on the number line along with their names.

The diagram below shows an example of students' jottings, where the numbers above the line show a connection with the numbers below the line. This opens the platform to discussing ordering fractions with different denominators too.

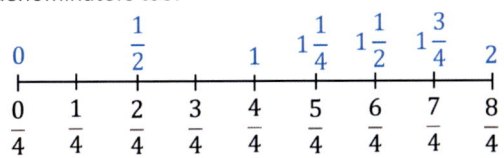

$$0 \quad \frac{1}{2} \quad 1 \quad 1\frac{1}{4} \quad 1\frac{1}{2} \quad 1\frac{3}{4} \quad 2$$

$$\frac{0}{4} \quad \frac{1}{4} \quad \frac{2}{4} \quad \frac{3}{4} \quad \frac{4}{4} \quad \frac{5}{4} \quad \frac{6}{4} \quad \frac{7}{4} \quad \frac{8}{4}$$

> **Notes:** Fraction Wall used in the examples are from MathsBot.com.
>
> An alternative manipulative that can be created is using strips of paper with equal height and equal length.
>
> If strips of paper are used, then the assumption here is that the student has a good understanding of constructing a fraction in relation to a whole as a concrete model. For example, the diagram below shows the manipulation with a strip of paper to show how a half can be created and connected to one whole. Let a strip of paper be called one whole (purple), then fold it into two equal parts (arrow from purple to green) then, open the fold to reveal two halves which are equal to one whole. 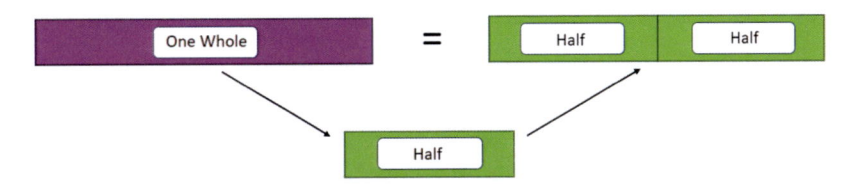
>
> One Whole = Half | Half
> Half
>
> Note that fractions in the form of the bar provide it to be seen as a discrete model, whereas fractions on a number line relate to the continuous model. The idea is the same as using 'number tracks' compared to 'number lines'.
>
> Also see FDP007 for connection with whole numbers.

> **FDP007 Statement:** If the numerator and the denominator are the same, then it is equal to one whole.
>
> **Answer:** ALWAYS TRUE
>
> **Manipulatives:** Fraction Wall and Number Line

This statement can be shown as always true through the use of a *Fraction wall* and a *Number Line*.

Assuming that we consider 1 whole and create fractions such that each part is equal then, we identify each part as a unit fraction of the 1 whole.

In the diagrams below, a half is created by physically manipulating 1 whole into two equal parts, similarly with a third being created by physically manipulating 1 whole into three equal parts and finally a quarter being created by physically manipulating 1 whole into four equal parts. The 1 whole has been shown in the diagram to enable the comparisons, conjectures, generalisations etc. As each construction was created from one whole, the understanding is that the total of all the parts would therefore be equal to one whole. This can be seen clearly through the use of a number line alongside the fractions.

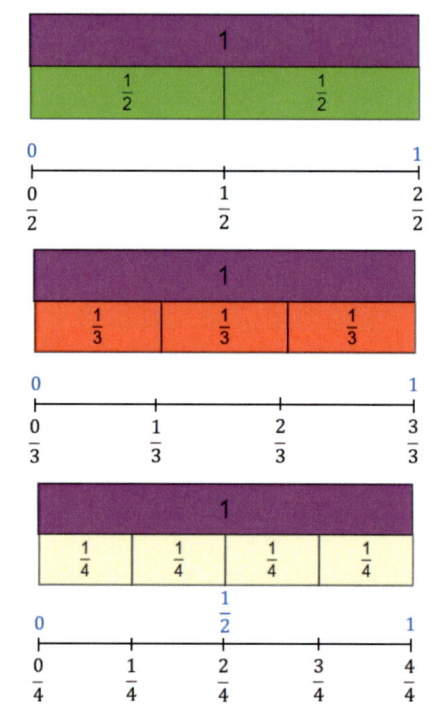

Encourage students to generalise from their findings and write statements whether in words, numbers, letters, diagrams or a combination to imply that any whole number divided by the same whole number is equal to one. For example: $\frac{2}{2} = 2 \div 2 = 1$.

Notes: Fraction Wall used in the examples are from MathsBot.com.

This investigation can be extended to other types of numbers (not just whole numbers), thus supporting the students to dive deeper into the conditions under which their generalisation is valid. For example, could they predict and show that any given quantity divided by the same quantity is equal to one.

Dividing a number by itself, n ÷ n = 1 or $\frac{n}{n} = 1$.

A note on creating your own fraction wall:

A fraction wall can be created by using strips of paper with equal height and equal length. Folding the strip of paper into equal parts will create unit fractions of the desired value.

If strips of paper are used, then the assumption here is that the student has a good understanding of constructing a fraction in relation to a whole as a concrete model. For example, the diagram below shows the manipulation with a strip of paper to show how a half can be created and connected to one whole. Let a strip of paper be called one whole (purple), then fold it into two equal parts (arrow from purple to green) then, open the fold to reveal two halves which are equal to one whole.

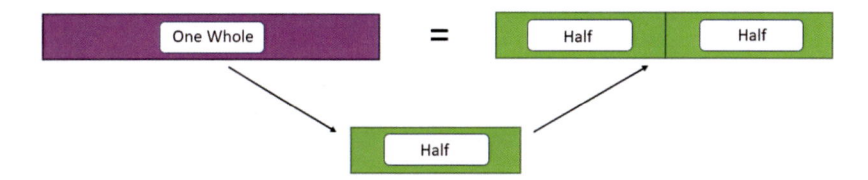

Note that fractions in the form of the bar provide it to be seen as a discrete model, whereas fractions on a number line relate to the continuous model. The idea is the same as using 'number tracks' compared to 'number lines'.

Also see FDP006 for connection with other fractions.

> **FDP008 Statement:** Fractions are less than one whole.
>
> **Answer:** SOMETIMES TRUE
>
> **Manipulatives:** Cuisenaire Rods and Number Lines

This statement can be shown as sometimes true through the use of *Cuisenaire rods* and *Number lines* under the condition that one whole is identified clearly from its relative equal fractional parts. For example, in the diagrams below, it can be seen irrespective of the lengths of A, B and C, that they have 3 equal parts (shown in the bars underneath them).

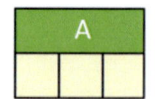

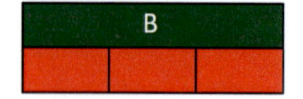

If the condition placed is that the length of B is 1 then, it can be represented using a number line showing fractional steps of thirds as seen in the diagram on the right.

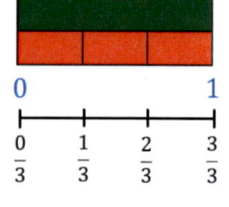

Hence, the statement is true for $\frac{0}{3}, \frac{1}{3}$ and $\frac{2}{3}$ as these are all less than 1. However, the statement is not true for $\frac{3}{3}$ as it is equal to 1.

If we extend the bars as in the diagram below and show a representation using the number line, then it can be shown that fractions exist beyond the value of one whole. Encourage students to draw the number lines alongside the Cuisenaire rods and label the number line. In the example provided, the number line has been labelled using mixed fractions as well as improper fractions. The diagram provides an opportunity to discuss how mixed numbers and improper fractions are connected.

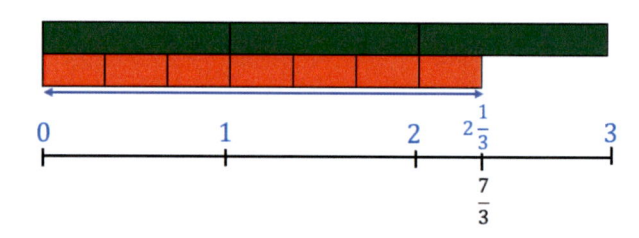

An improper fraction is one where the numerator is greater than the denominator, as in $\frac{7}{3}$ above and will therefore be greater than one whole.

In conclusion, this statement is sometimes true. Encourage students to investigate other fractions to generalise their findings. Discussions about how a division sentence can be represented as a fraction will support connecting with improper fractions. For example, using the example above: $\frac{7}{3} = 7 \div 3$. Could this be true for all fractions?

> **Notes:** Cuisenaire Rods used in the examples are from MathsBot.com.

> **FDP009 Statement:** Doubling both the numerator and denominator creates a new fraction that is double in size to the original one.
>
> **Answer:** NEVER TRUE
>
> **Manipulatives:** Fraction Wall

This statement can be shown as never true through the use of a *Fraction wall* and the assumption that the student has sound understanding of the idea of doubling with fractions. Encourage discussions about how doubling relates to multiplying by 2 as well as repeated addition.

A systematic approach is used in investigating the statement to enable generalising for different types of fractions.

First let's consider a unit fraction: $\frac{1}{2}$.
Doubling both the numerator and denominator can be seen in the following calculation:
$$\frac{1}{2} \times \frac{2}{2} = \frac{1 \times 2}{2 \times 2} = \frac{2}{4} = \frac{1}{2}$$

However, this shows that the unit fraction $\frac{1}{2}$ has remained unchanged. Discuss with students about the impact of multiplying the numerator and denominator by the same value and how it connects with equivalent fractions.

The diagram below shows the act of doubling one half by joining two halves together horizontally and the jottings along with the arrow indicates the size of the resulting fraction.

$$\frac{1}{2} \times 2 = \frac{2}{2} = 1 \quad \text{or} \quad \frac{1}{2} + \frac{1}{2} = \frac{2}{2} = 1$$

Note that students may show the jottings of $\frac{1}{2} \times 2 = \frac{2}{2} = 1$ as $\frac{1}{2} \times 2 = \frac{1}{2} \times \frac{2}{1} = \frac{1 \times 2}{2 \times 1} = \frac{2}{2} = 1$.

From the above jottings the statement is not true as both calculations reveal different resulting values. To support generalising for all unit fractions, encourage students to consider investigating several other unit fractions too.

Now, let's consider a proper fraction: $\frac{3}{4}$

Doubling both the numerator and denominator can be seen in the following calculation:
$\frac{3}{4} \times \frac{2}{2} = \frac{6}{8} = \frac{3}{4}$ However, this shows that the fraction $\frac{3}{4}$ remains unchanged.

The diagram on the following page shows the act of doubling three quarters (or three fourths) by joining two of them together horizontally and the jottings along with the arrow indicates the size of the resulting fraction. Students can compare their calculations to the fraction wall bars by placing the 1 whole and 1 half bars as shown.

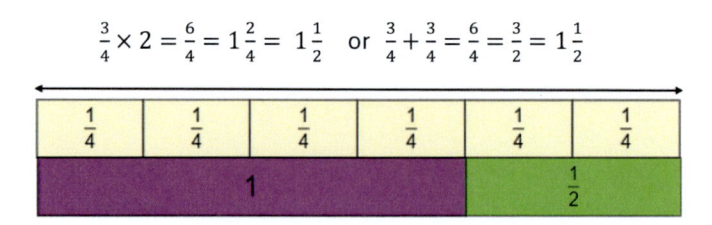

$$\tfrac{3}{4} \times 2 = \tfrac{6}{4} = 1\tfrac{2}{4} = 1\tfrac{1}{2} \quad \text{or} \quad \tfrac{3}{4} + \tfrac{3}{4} = \tfrac{6}{4} = \tfrac{3}{2} = 1\tfrac{1}{2}$$

Again the same findings appear to be the case. The statement is not true. Students should be encouraged to investigate further examples of proper fractions, discuss their findings and make conjectures and generalisations.

Now, let's consider a whole number: $1 = \tfrac{1}{1}$. Here, we are considering whole numbers as they can be written in fractional form.

Doubling both the numerator and denominator can be seen in the following calculation:
$\tfrac{1}{1} \times \tfrac{2}{2} = \tfrac{2}{2} = \tfrac{1}{1}$ However, this shows that the fraction $\tfrac{1}{1}$ remains unchanged.

The diagram below shows the act of doubling 1 by joining two of them together horizontally and the jottings along with the arrow indicates the size of the resulting fraction, in this case a whole number (2).

$$\tfrac{1}{1} \times 2 = \tfrac{2}{1} = 2 \quad \text{or} \quad \tfrac{1}{1} + \tfrac{1}{1} = \tfrac{1+1}{1} = \tfrac{2}{1} = 2$$

This supports the statement to be not true. Students should be encouraged to investigate further examples of whole number written in their fractional form and discuss their findings.

Lastly, let's consider an improper fraction $\tfrac{3}{2}$.

Doubling both the numerator and denominator can be seen in the following calculation:
$\tfrac{3}{2} \times \tfrac{2}{2} = \tfrac{6}{4} = \tfrac{3}{2}$ However, this shows that the fraction $\tfrac{3}{2}$ remains unchanged.

The diagram below shows the act of doubling three halves (or one and a half) by joining two of them together horizontally and the jottings along with the arrow indicate the size of the resulting fraction. Students can compare their calculations to the fraction wall bars as seen in the diagram below.

$$\tfrac{3}{2} \times 2 = \tfrac{6}{2} = 3 \quad \text{or} \quad \tfrac{3}{2} + \tfrac{3}{2} = \tfrac{6}{2} = \tfrac{3}{1} = 3$$

Again, the same findings appear to be the case. The statement is not true.

Students should be encouraged to investigate further examples of improper fractions, discuss their findings and generalise.

> **Notes:** Fraction Wall used in the examples is from MathsBot.com.
>
> Note that Cuisenaire rods can be used in exactly the same way as the fraction wall however, as the rods are not assigned a particular value, students will need to ensure that a value is assigned per fraction that is investigated.
>
> A common misconception when comparing fractions is that the fraction with the largest denominator is the largest fraction - see FDP001.
>
> Another common misconception when adding fractions is that the numerators are added and so are the denominators, possibility of it stemming from the rule associated with multiplying fractions.

FDP010 Statement: Dividing a fraction by 2 is the same as multiplying it by $\frac{1}{2}$.

Answer: ALWAYS TRUE

Manipulatives: Fraction Wall

This statement can be shown as always true through the use of a *Fraction wall* and by exploring different types of fractions, initially with some unit fractions.

Encourage students to begin with some familiar unit fractions such as $\frac{1}{2}, \frac{1}{3}, \frac{1}{4}$ and so on. Beginning with considering $\frac{1}{2}$:

Using this wall, we can see that a half divided into two equal parts is $\frac{1}{4}$. So $\frac{1}{2} \div 2 = \frac{1}{4}$.

Students may also see it like this:

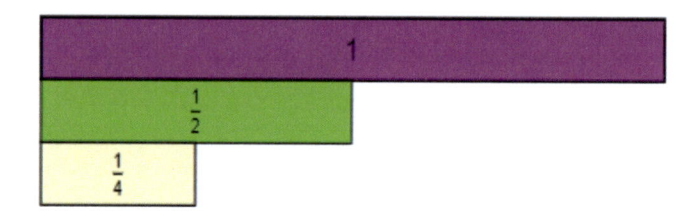

Students could use Cuisenaire rods to manipulate this and check that two quarters are the same length as a half. The rods they choose will depend on what they choose as their '1' whole.

We also know from our rules about multiplying fractions, that this is done by multiplying the numerators together to form the new numerator and multiplying the denominators together to form the new denominator.

$\frac{1}{2} \times \frac{1}{2} = \frac{1 \times 1}{2 \times 2} = \frac{1}{4}$ So, multiplying $\frac{1}{2}$ by $\frac{1}{2}$ also results in $\frac{1}{4}$.

Hence, the statement would be true.

Next, we might consider $\frac{1}{3}$:

Using this wall, we can see that a third divided into two 2 equal parts is $\frac{1}{6}$. So $\frac{1}{3} \div 2 = \frac{1}{6}$

Again, students may see it like this:

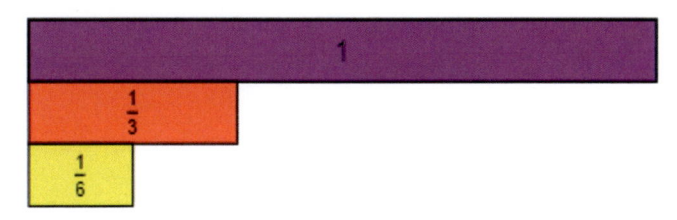

Students could use Cuisenaire rods to manipulate this and check that two sixths are the same length as a third, again the rods chosen will depend on what is chosen to represent '1' whole.

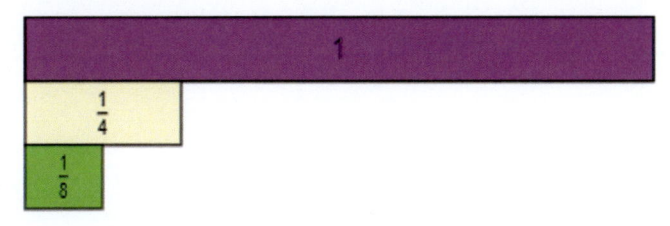

Working out a third multiplied by a half:

$\frac{1}{3} \times \frac{1}{2} = \frac{1 \times 1}{3 \times 2} = \frac{1}{6}$ So, multiplying $\frac{1}{3}$ by $\frac{1}{2}$ also results in $\frac{1}{6}$.

Hence, the statement would be true.

Similarly, dividing a quarter into two equal parts results in an eighth as shown in the diagram here.

Again, students may see it like this:

Working out a quarter multiplied by a half:

$\frac{1}{4} \times \frac{1}{2} = \frac{1 \times 1}{4 \times 2} = \frac{1}{8}$ So, multiplying $\frac{1}{4}$ by $\frac{1}{2}$ is equal to $\frac{1}{8}$.

Hence, the statement would be true.

Students could try to show this visually with other fractions such as non-unit fractions but numerators that are multiples of 2 would be useful! Consider beginning with $\frac{2}{3}$.

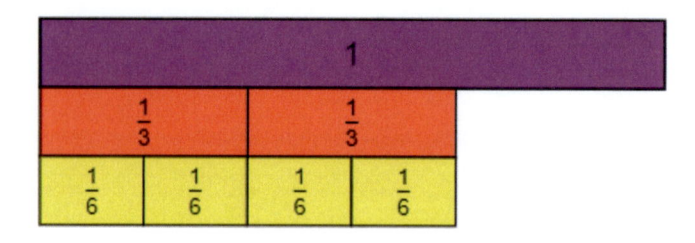

So, $\frac{2}{3} \div 2$ would give $\frac{2}{6} = \frac{1}{3}$ (they are equivalent).

Students may see it like this:

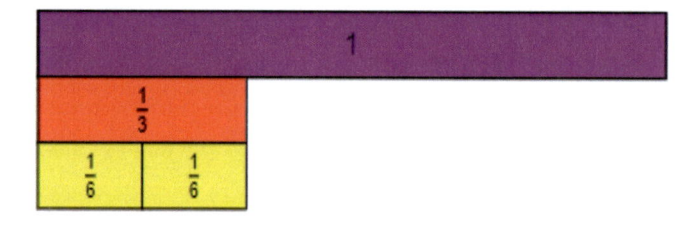

Working out two-thirds multiplied by a half:

$\frac{2}{3} \times \frac{1}{2} = \frac{2 \times 1}{3 \times 2} = \frac{2}{6}$ So, multiplying $\frac{2}{3}$ by $\frac{1}{2}$ also equates to $\frac{1}{3}$.

Hence, the statement would still be true.

Encourage students to conjecture the inverse relationship between doubling and halving with different types of numbers, leading to a generalisation.

Notes: Fraction Wall in the examples are from MathsBot.com.

A unit fraction is one with a numerator of 1.

Using an array approach may also be useful. Paper and bar drawings can be used in the following way:

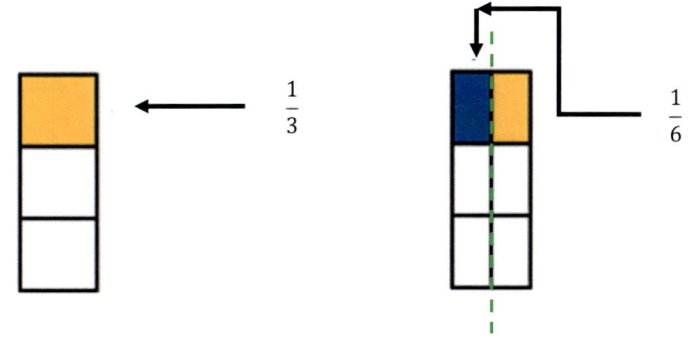

The shaded part is a third, then when this shape is divided in two with a vertical line (shown with the dashed line), it has been halved and the darker shaded part therefore shows half of a third, which is a sixth. It is important to show the 'whole array' in order to relate the fractional amount to it.

Students might also approach this through the idea of symmetry, to support the inverse relationship between doubling and halving. They should come to the generalisation that multiplying by a half is the same as dividing by two, multiplying by a third is the same as dividing by three and so on.

This statement could be investigated using different types of fractions besides unit fractions and proper fractions. The principal idea of understanding size of a fraction using a common denominator remains the same.

FDP011 Statement: 10% of a number is the same as the number multiplied by 0.1.

Answer: ALWAYS TRUE

Manipulatives: Hundred Square and Bar Modelling

This statement can be shown as always true through the use of a *Hundred square and Bar modelling*. Depending on where your students are, you may want to begin investigating with whole numbers before moving onto decimals, fractions or negative numbers for example.

Encourage students to think about the notation and what percentage means. It means "out of 100". So, a blank hundred square is a useful starting point.

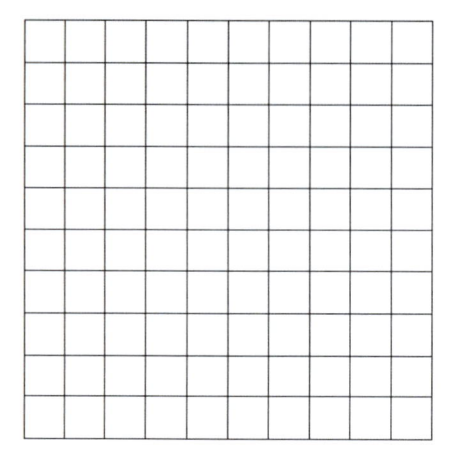

Now 10% means 10 parts out of a 100 (equal parts). Hence, a hundred square is helpful as 10% can be shown as 10 squares on the hundred square (shown in a different shade below).

If we consider finding 10% of 300 as an example, then each square is worth 3 and 10% will be 3 multiplied by 10 which is 30. Clearly, this approach can be used with numbers other than 300.

Now if we want to check how to multiply by 0.1, we should understand what 0.1 means. This is a '1' in the tenths place – so 0.1 can also be written as $\frac{1}{10}$. Hence, multiplying by 0.1 can also be interpreted as dividing by 10 or finding a tenth.

A tenth of 300 can be shown on a bar model that has been divided into 10 equal parts, where the whole bar represents 300. See diagram below.

From this, 300 divided into 10 equal parts is 30 so one tenth of 300, or 300 x 0.1 = 30.

This is the same as finding 10% of 300 so the statement is true.

Turning to a more abstract approach and using algebra for a general number n:
10% of a number n is $\frac{10}{100} \times n = \frac{10n}{100} = \frac{n}{10}$ and $0.1 \times n = \frac{1}{10} \times n = \frac{n}{10}$.

Hence, the statement is always true.

Notes: Hundred Square and Bar Model in the examples are from MathsBot.com.

Percentage means "out of 100".

Students should understand how to multiply by 0.1 and be able to recognise this as a tenth.

> **FDP012 Statement:** Finding 5% of a number is the same as dividing the number by 5.
>
> **Answer:** NEVER TRUE
>
> **Manipulatives:** Hundred Square

This statement can be shown as never true through the use of a *Hundred square* and by comparing 5% of a number with dividing the number by 5.

Students might jump to conclusions about 5% being the same as dividing by 5. This could suggest that they don't understand what the percentage notation actually means. For example: 5% means $\frac{5}{100}$.

Encourage students to use integers to start with, for example, 200. We have chosen a multiple of 100 as this is taught prior to learning about division with decimal quotients. For example, 200 divided by 100 is equal to a whole number quotient.

The number 200 can be represented by the hundred square such that, each square must be equal to 200 ÷ 100 = 2.

So, 5% is 5 of the squares and 5 x 2 = 10.

It is shown here with 5 shaded squares representing the 5%.

We can also calculate 5% of 200 using the multiplication method. This can be seen here:

$$\frac{5}{100} \times \frac{200}{1} = \frac{5 \times 200}{100 \times 1} = \frac{1000}{100} = 10.$$

Students may also calculate this in different ways such as:

$\frac{5}{100} \times 200 = \frac{5}{10} \times 20 = \frac{100}{10} = 10$ or $\frac{5}{100} \times 200 = \frac{1}{20} \times 200 = \frac{200}{20} = \frac{20}{2} = 10$.

While 200 ÷ 5 = 40 and is therefore not equal to 10 making the statement false.

Now using the number 60, this can be represented by the hundred square such that, each square must be equal to: 60 ÷ 100 = 0.6.

So, 5% is 5 of the squares and 5 x 0.6 = 3.0 (discuss with the students whether it is essential to write 3.0 or whether 3 would suffice).

0.6	0.6	0.6	0.6	0.6					

Calculating 5% of 60 using the multiplication method: $\frac{5}{100} \times \frac{60}{1} = \frac{5 \times 60}{100 \times 1} = \frac{300}{100} = 3$.

Students may also calculate this in different ways such as:

$\frac{5}{100} \times 60 = \frac{5}{10} \times 6 = \frac{30}{10} = 3$ or $\frac{5}{100} \times 60 = \frac{1}{20} \times 60 = \frac{60}{20} = 3$.

What happens when we divide our chosen number by 5? Now, 60 ÷ 5 = 12.
Hence, the statement is false for the case of 60.

Exploring the statement with a decimal number, say 0.5, equates each square to be worth 0.005 (0.5 ÷ 100 = 0.005).

0.005	0.005	0.005	0.005	0.005					

Five of the squares is therefore worth 0.005 x 5 = 0.025.

Again, calculating 5% of 0.5 using the multiplication method:

$$\frac{5}{100} \times 0.5 = \frac{5}{100} \times \frac{5}{10} = \frac{25}{1000} = 0.025.$$

What happens when we divide our chosen number by 5? 0.5 ÷ 5 = 0.1.

Clearly, the statement is not true again as 0.025 is not equal to 0.1.

Notes: The Hundred Square used in the examples is from MathsBot.com.

Students can investigate other connections with percentage of amounts and their division counterpart. Refer to FDP010 and FDP011.

> **FDP013 Statement:** 30% of 80 is the same as 80% of 30.
>
> **Answer:** ALWAYS TRUE (for any numbers used)
>
> **Manipulatives:** Hundred Square

This statement can be shown as always true through the use of a *Hundred square*.

In the example of 30% of 80, 30% means 30 out of 100. This has been shown on the square below with the 30 shaded squares. Now 1 percent is 1 out of 100 so in a square that's representing our 'whole', in this case 80, each unit square is worth 80 ÷ 100 = 0.8.

Hence, 30% of 80 is worth 30 squares of 0.8 or 30 lots of 0.8 or 0.8 x 30 = 24.

Now, let's consider what 80% of 30 might look like. 80% means 80 out of 100 which has been shown here with the 80 shaded squares.

As said above, 1 percent is 1 out of 100 so in a square that's representing our 'whole', in this case 30, each unit square is: 30 ÷ 100 = 0.3.

Hence, 80% of 30 is worth 80 squares of 0.3 or 80 lots of 0.3 or 0.3 x 80 = 24.

When comparing the two diagrams it is not easy to see that the results are the same. However, the calculations show 30% of 80 is the same as 80% of 30 as they are both 24.

Encourage students to try further examples, such as 40% of 50 and 50% of 40 or 10% of 20 and 20% of 10 and so on.

Encourage students to make a generalisation in words that the statement is valid for all whole numbers or indeed for any type of number.

> **Notes:**
>
> This statement can be looked at with the algebra once students have considered several examples.
>
> x% of y can be written as:
> $$\frac{x}{100} \times y = \frac{xy}{100}$$
>
> While y% of x can be written as:
> $$\frac{y}{100} \times x = \frac{yx}{100}$$
>
> Because multiplication is commutative this means that xy is equal to yx and the two amounts are the same.

> **FDP014 Statement:** Reducing a whole number by 10% followed by increasing it by 10% results in returning to the original number.
>
> **Answer:** NEVER TRUE
>
> **Manipulatives:** Hundred Square

This statement can be shown as never true through the use of a *Hundred square*.

Let's begin by considering 10% of 80. 10% means 10 out of 100. This has been shown on the square below with the 10 shaded squares. Now 1 percent
is 1 out of 100 so in a square that's representing our 'whole', in this case 80, each unit square is worth 80 ÷ 100 = 0.8 (links with FDP013).
Hence, 10% of 80 is worth 10 squares of 0.8 or 0.8 x 10 = 8.

0.8	0.8	0.8	0.8	0.8	0.8	0.8	0.8	0.8	0.8

Now, reducing this number, 80, by 10% means subtracting 8 from 80. So, 80 – 8 = 72.

Now, increasing this (new) number by 10% means finding 10% of 72 and adding it onto 72. However, we know that 10% of 80 is 8 and that's what we need to increase the 72 by to get back to 80, but our new number is 72 (not 80) and 10% of 72 is 7.2 and not 8.

Let's consider another example: 150.
Our ten shaded squares on the following page each show 1.5 as 1% is worth 1.5 and ten of these is:
10 x 1.5 = 15.
So 10% of 150 is 15.

1.5	1.5	1.5	1.5	1.5	1.5	1.5	1.5	1.5	1.5

Now, reducing this number, 150, by 10% means subtracting 15 from 150. So, 150 − 15 = 135.

Now, increasing this (new) number by 10% means finding 10% of 135 and adding it onto 135. However, we know that 10% of 150 is 15 and that's what we need to increase the 135 by to get back to 150, but our new number is 135 (not 150) and 10% of 135 is 13.5 and not 15.

The diagrams support the understanding of 10% of a number. However, the calculations show that the answers are not the same, therefore the statement is still not true.

Encourage students to generalise their findings and try other examples until they have convinced themselves and their peers.

Notes: Hundred Square used in the example is from MathsBot.com.

Using an algebraic notation, a 10% decrease is the same as finding 90% of a number n:

$$\frac{90}{100} \times n = 0.9n$$

Now, increasing this result by 10% means finding 110% of it:

$$\frac{110}{100} \times 0.9n = 0.99n$$

As 0.9 of n is not the same as 0.99 of n – the statement will never be true.

FDP015 Statement: A quarter (1/4) can be written as 1.4.

Answer: NEVER TRUE

Manipulatives: Fraction Wall

This statement can be shown as never true through the use of a *Fraction wall* by comparing a quarter with the decimal number 1.4.

The statement is comparing two different types of numbers, a fraction and a decimal. As a result, students should be familiar with converting fractions into tenths, hundredths and thousandths (powers of tenths).

Encourage students to think what a quarter means and to find different ways to show it. It is shown below using a fraction wall.

A quarter represents one part when the whole is divided into four equal parts.

Now, encourage students to think how they might show 1.4. What does this decimal number mean? It means 1 whole and 4 tenths. Since, $\frac{1}{10} = 0.1$ then, $\frac{4}{10} = 0.4$.

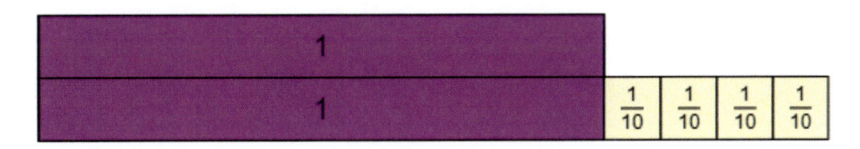

Comparing a quarter with 1.4 we can see they are not the same thing at all!

Students might also want to find how to write a quarter as a decimal. Encourage them to use their knowledge of equivalent fractions and express this out of 100.

$$\frac{1}{4} \times \frac{25}{25} = \frac{25}{100}$$

Now, they can use their knowledge of division or decimals to show that 25 hundredths are 20 hundredths and 5 hundredths or $\frac{25}{100} = \frac{20}{100} + \frac{5}{100}$, which is equivalent to $\frac{2}{10} + \frac{5}{100}$.

This can also be represented in decimal form as $25 \div 100 = 0.25$.

> **Notes:** Fraction Walls in the examples are from MathsBot.com.
>
> Students sometimes jump to generalisations with fractions and decimals and might think that 1.4 is the same as the fraction $\frac{1}{4}$ because they see the digits 1 and 4. Discuss the comparison with $\frac{1}{2}$ being equivalent to 0.5 rather than 1.2.
>
> A further investigation can be to explore if there are decimals that do use the same digits to form a fraction.

> **FDP016 Statement:** When comparing decimals, the longer the decimal the bigger the number.
>
> **Answer:** SOMETIMES TRUE
>
> **Manipulatives:** Dienes

This statement can be shown as sometimes true through the use of *Dienes* and by comparing some decimals that students can make with the understanding of the representation given in the diagram below. By 'longer', we mean that it has more digits written. So, for example, 4.523 has four digits and 4.5 has two digits and therefore, 4.523 is 'longer'.

Using Dienes to represent decimals is defined as below.

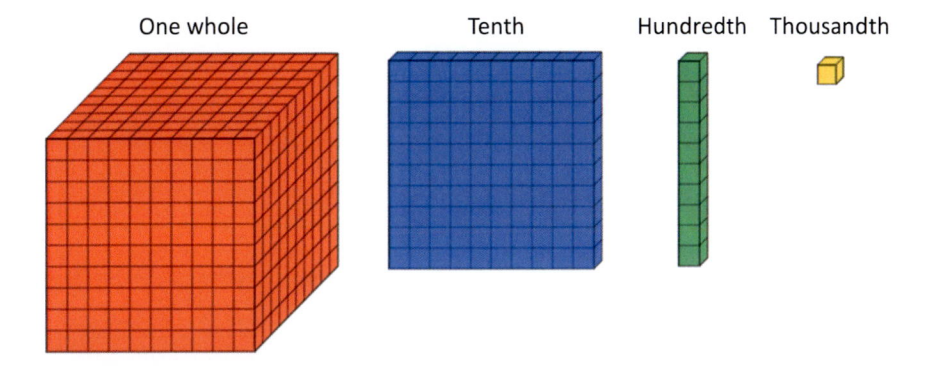

One whole   Tenth   Hundredth   Thousandth

These can now be used to represent decimals numbers, such as 1.32 as shown below.

Ones	Tenths	Hundredths	Thousandths

Similarly, 1.237 can be represented using the Dienes as shown below.

Ones	Tenths	Hundredths	Thousandths

When comparing these two numbers, start by comparing the largest place value then, the second largest place value, and so on. This clearly shows that 1.32 is greater than 1.237 as the tenths place value is greater in the number 1.32 than the number 1.237. This means 1.32 > 1.237.
However, the 'longer' decimal is 1.237.
This example supports the statement as **not** being true.

However, comparing the numbers 1.125 and 1.12 (see diagrams below), it can be shown that 1.125 is 5 thousandths greater than 1.12. Students could use Dienes to convince themselves and others. For this example, the statement is true with the 'longer' decimal being the greater one.

Ones	Tenths	Hundredths	Thousandths

Ones	Tenths	Hundredths	Thousandths

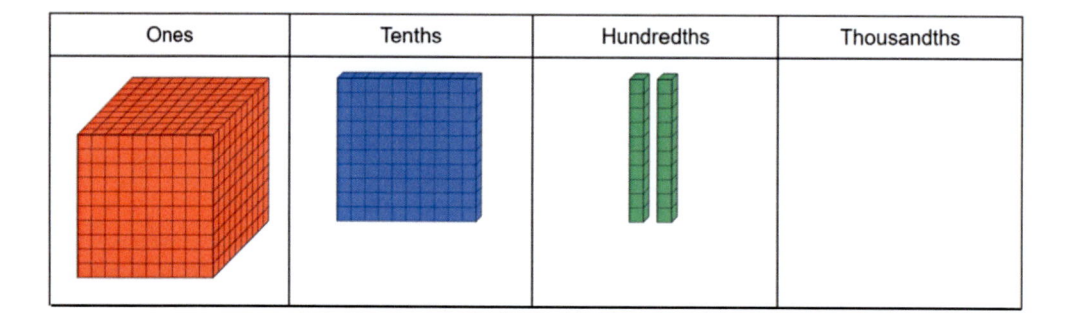

Hence, the statement overall is sometimes true.

Discuss with students the lack of value in the thousandths place. How might this be represented symbolically? Think of the idea of zero as a valid value particularly for the purpose of comparison. Students may jot their decimal numbers to compare by aligning the digits in the same place values hence employing the comparison of size by place value to support their findings.

> **Notes:** Dienes in the examples are from MathsBot.com.
>
> A hundred bead string can be used to support the comparison of decimal numbers from 0.01 to 1.00.
>
> Similarly, a representation of a number line such as a metre stick or measuring tape can be used.

> **FDP017 Statement:** Decimal numbers can be written as fractions.
>
> **Answer:** ALWAYS TRUE
>
> **Manipulatives:** Dienes

This statement can be shown as always true through the use of *Dienes* and by appreciation of the place value system.

Let us define the Dienes to be these values:

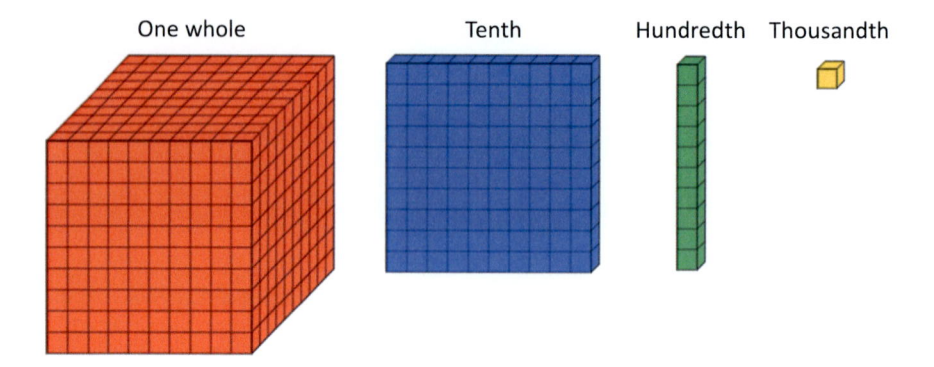

One whole   Tenth   Hundredth   Thousandth

Let's consider numbers smaller than 1. For example, 0.3 (beginning with one decimal place). This can be represented like this:

Ones	Tenths	Hundredths

These three 'squares' represent 0.3 of the whole (the large 'red' cube). As we have a 3 in the tenths place, this can be written as a fraction: $\frac{3}{10}$. Hence 0.3 = $\frac{3}{10}$.

Another example, this time with 2 decimal places, is considered: 0.25.

This can be made as shown below with two of the blue 'squares' in the tenths place, and five of the green rods in the hundredths place.

Ones	Tenths	Hundredths

This decimal can be written as 0.25 which can then be written as $\frac{2}{10} + \frac{5}{100} = \frac{20}{100} + \frac{5}{100} = \frac{25}{100}$.

This fraction can be simplified to $\frac{25 \div 25}{100 \div 25} = \frac{1}{4}$.

This brings the opportunity to discuss equivalent fractions with the students. As well as, why $\frac{1}{4}$ is not equal to 0.14 or 1.4 (See FDP015).

Finally, let's consider a number greater than 1. For example, 1.125 as shown here:

Ones	Tenths	Hundredths	Thousandths

Now this can be written as $1 + \frac{1}{10} + \frac{2}{100} + \frac{5}{1000}$

or $\frac{1000}{1000} + \frac{100}{1000} + \frac{20}{1000} + \frac{5}{1000} = \frac{1125}{1000}$

or $1 + \frac{100}{1000} + \frac{20}{1000} + \frac{5}{1000} = 1\frac{125}{1000}$, which can be simplified further to $1\frac{1}{8}$.

Encourage students to generalise that all decimals can be written as fractions.

> **Notes:** Dienes in the examples are from MathsBot.com.
>
> Students should be aware of the position of the decimal point in relation to place value. This is assumed with the naming of the place value columns in the diagrams used in the examples.
>
> Other investigations that can be pursued are:
> Is the reverse true or not? (Fractions can be converted into decimal numbers.); Decimal numbers can be written as percentages and vice versa.

> **FDP018 Statement:** Fractions can be simplified.
>
> **Answer:** SOMETIMES TRUE
>
> **Manipulatives:** Fraction Wall

This statement can be shown as sometimes true through the use of a *Fraction wall* by comparing fractions.

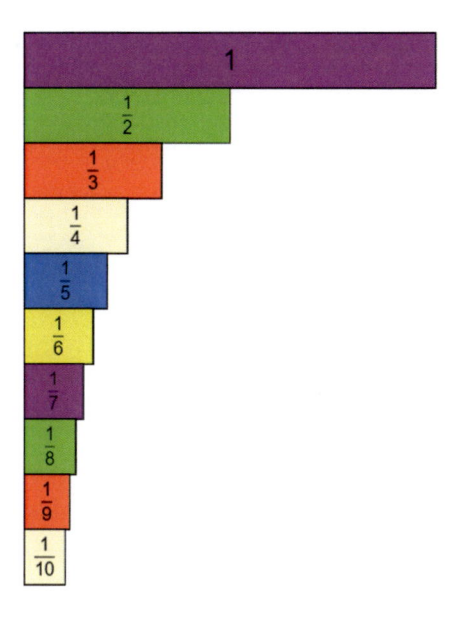

First of all, let's consider unit fractions. The diagram below shows unit fractions to a tenth.

When looking at unit fractions, the numerator is always 1 – by definition. These fractions cannot be simplified further as the highest common factor of the numerator and denominator is 1. As dividing the numerator and denominator by 1 does not change the value of the digits, we can state that the fraction is in its simplest form. This would indicate the statement to be not true.

Now consider $\frac{2}{4}$ or $\frac{3}{6}$, these can be explored through examining a fraction wall image such as the one shown below.

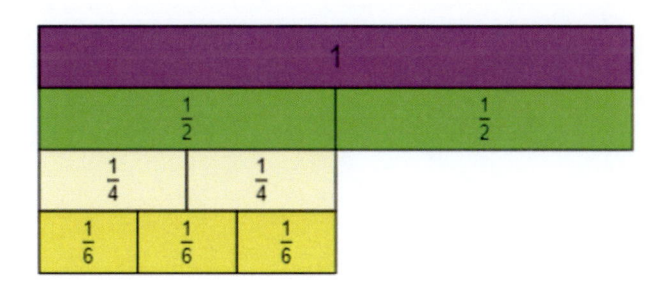

This shows that the fractions $\frac{2}{4}$ or $\frac{3}{6}$ can both be simplified to $\frac{1}{2}$. Indicating the statement to be true.

Other fractions can also be considered such as $\frac{4}{6}$ (shown below) or $\frac{40}{60}$ which can be simplified to $\frac{2}{3}$. (See FDP009)

Encourage students to consider improper fractions before they generalise their findings for all fractions or for different types of fractions. For example, a student may articulate that a fraction can be simplified if the numerator (top number) and the denominator (bottom number) can both be divided by a common factor greater than 1.

> **Notes:** Fraction Wall in the examples are from MathsBot.com.
>
> Cuisenaire rods can be used as an alternative manipulative.
>
> A numerator is the number above the vinculum.
> A denominator is the number below the vinculum.
> A proper fraction is a fraction where the numerator is smaller than the denominator.
> An improper fraction is a fraction where the numerator is greater than the denominator.
> A vinculum is the fraction bar (horizontal line between the numerator and denominator).

Chapter 4

Geometry and Shape Statements

A triangle has three acute angles. GS001	A trapezium has two sides of equal length. GS008
An equilateral triangle has all sides the same length, but the angles can be different. GS002	Any quadrilateral can be made from two triangles. GS009
A triangle has perpendicular sides. GS003	The interior angles of a pentagon total 360 degrees. GS010
An equilateral triangle has one line of symmetry. GS004	A pentagon has two right angles. GS011
A square is a rectangle. GS005	Hexagons have sides that are equal in length. GS012
A kite is a rhombus. GS006	The circumference is approximately three times the diameter. GS013
A square has four lines of symmetry. GS007	If the area of a rectangle is 24 squared centimetres, then the side lengths are 4cm and 6cm. GS014

The perimeter of a rectangle is four times one of the sides. GS015	A pyramid has an even number of faces. GS019
The area of a triangle is $\frac{1}{2}$ x base x height. GS016	A prism has at least three rectangular faces. GS020
Doubling the width of a rectangle will result in the area being doubled. GS017	A regular polygon will tessellate with itself. GS021
Two rectangles have the same perimeter, so they will have the same area. GS018	

> **GS001 Statement:** A triangle has three acute angles.
>
> **Answer:** SOMETIMES TRUE
>
> **Manipulatives:** Triangles

This statement can be shown as sometimes true through the use of different size and shaped *Triangles* with varying angles.

Before investigating the statement, it is important to ensure that the students understand the definition of an acute angle. An interpretation can be that each angle measured or otherwise needs to be identified as an acute angle or not an acute angle.

This statement may be investigated through two different approaches: using angle rules and properties of triangles or measuring angles inside a variety of triangles.

In the information below, the approach employed is by investigating the statement through measuring varying angles for different categories of triangles whilst keeping the labelling of the vertices constant. Encourage students to record their measurements in a systematic way that promotes noticing any emerging patterns.

Encourage students to discuss whether the angle under question is acute or not, using the definition of an acute angle (0° < acute angle < 90°).

Begin with measuring the angles in an equilateral triangle, as seen in the diagram. In this particular case, students may state that due to the property of an equilateral triangle, there is no need to measure all three angles in an equilateral triangle.

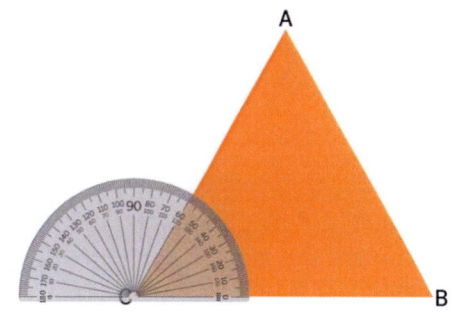

Next, students may investigate right-angled triangles which can be isosceles and scalene, as seen in the diagrams below. As students record their findings, they may use their prior knowledge such as right angles are not acute angles, therefore, the statement is not true in this case.

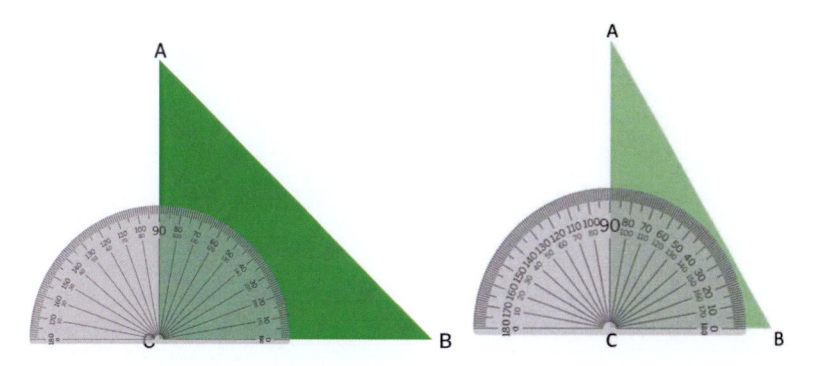

Finally, students may investigate scalene triangles.

Scalene Triangle 1:

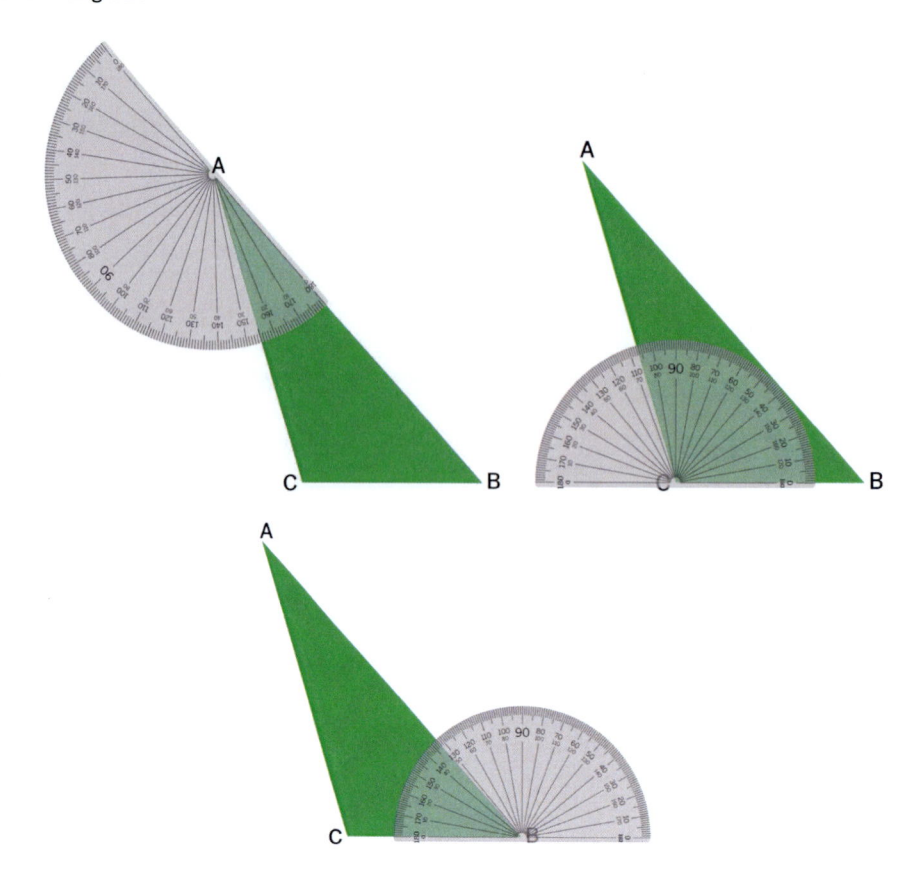

These show that two of the angles are acute and one is not an acute angle. For this case, the statement is not true.

Scalene Triangle 2:

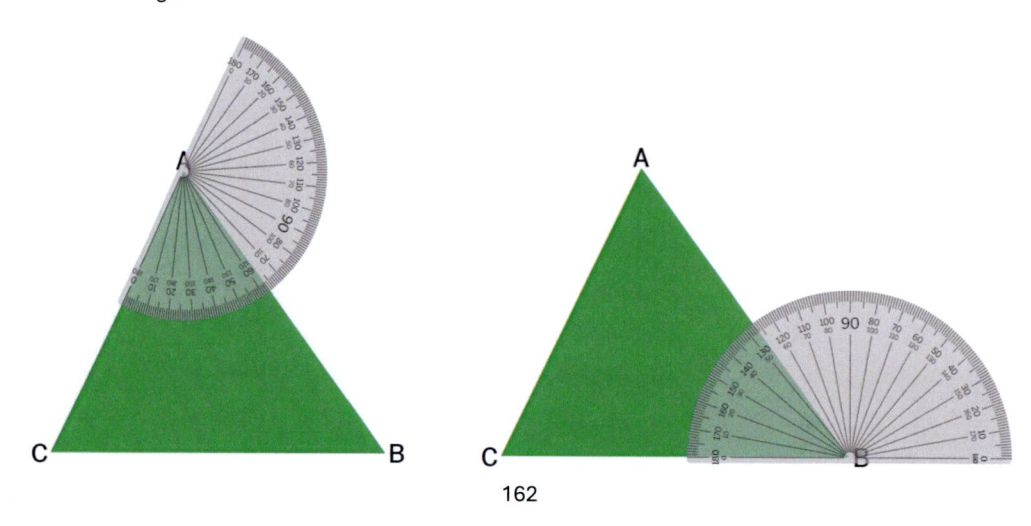

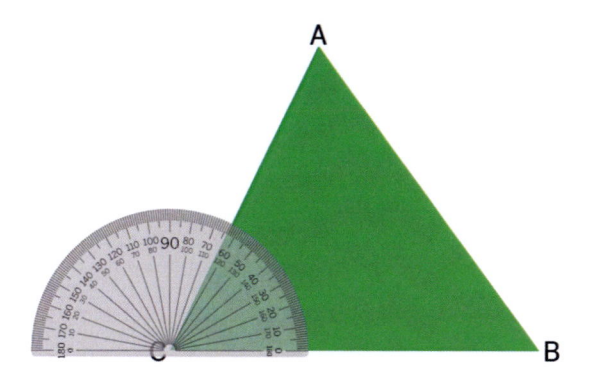

These show that all of the angles are acute. So, for this case, the statement is true.

From the examples above, the statement appears to be sometimes true.

A suggestion of the recordings for the examples above could be tabulated as shown below:

Type of Triangle	Angle A	Angle B	Angle C
Equilateral	60° - acute	60° - acute	60° - acute
Right-angled			90° - not acute
Scalene Triangle 1	25° - acute	48° - acute	107° - not acute
Scalene Triangle 2	63° - acute	53° - acute	64° - acute

Although, the above sample is small, students may predict that the statement is sometimes true and continue to investigate under specified conditions. For example, whilst investigating isosceles triangles, they may create a table of values of angles to support them in making generalisations where it is approached in a systematic way without the use of measuring and through knowledge of the properties of an isosceles triangle. The jottings and calculations in the table below show a possible approach.

Angle A	Angle B	Angle C
1°	89.5°	89.5°
2°	89°	89°
3°	88.5°	88.5°
4°	88°	88°

Angle A	Angle B	Angle C
91°	44.5°	44.5°
92°	44°	44°
93°	43.5°	43.5°
94°	43°	43°

Angle A	Angle B	Angle C
176°	2°	2°
177°	1.5°	1.5°
178°	1°	1°
179°	0.5°	0.5°

From their findings, students may infer that the statement is sometimes true for specific types of triangles.

> **Notes:** Triangles in the examples are from Polypad.org.
>
> This activity can be done by cutting triangular shapes from paper and measuring their respective interior angles.
>
> Statement GS003 can be referred to for further investigation and connection with angles in a triangle.

An alternative approach is to use a set square instead of a protractor as the measuring tool to identify if an angle is acute or not acute through the comparison of the right-angle in a set square, as seen in the diagram below.

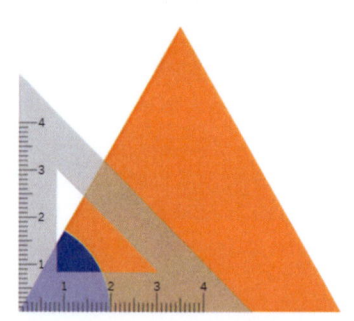

Students can further investigate the statement 'A triangle can have two obtuse angles'.

> **GS002 Statement:** An equilateral triangle has all sides the same length, but the angles can be different.
>
> **Answer:** NEVER TRUE
>
> **Manipulatives:** Equilateral Triangles

This statement can be shown as never true through the use of different side length *Equilateral Triangles*.

Before investigating the statement, it is important to ensure that the students are able to measure side lengths of any triangle and are able to use a protractor/angle measurer to measure the interior angles of a triangle.

Let's consider three different sized triangles. The diagrams below show equilateral triangles with lengths 4cm, 4.7cm and 5.4cm showing interior angle measurement of 60°. Students may measure all the interior angles and record their findings.

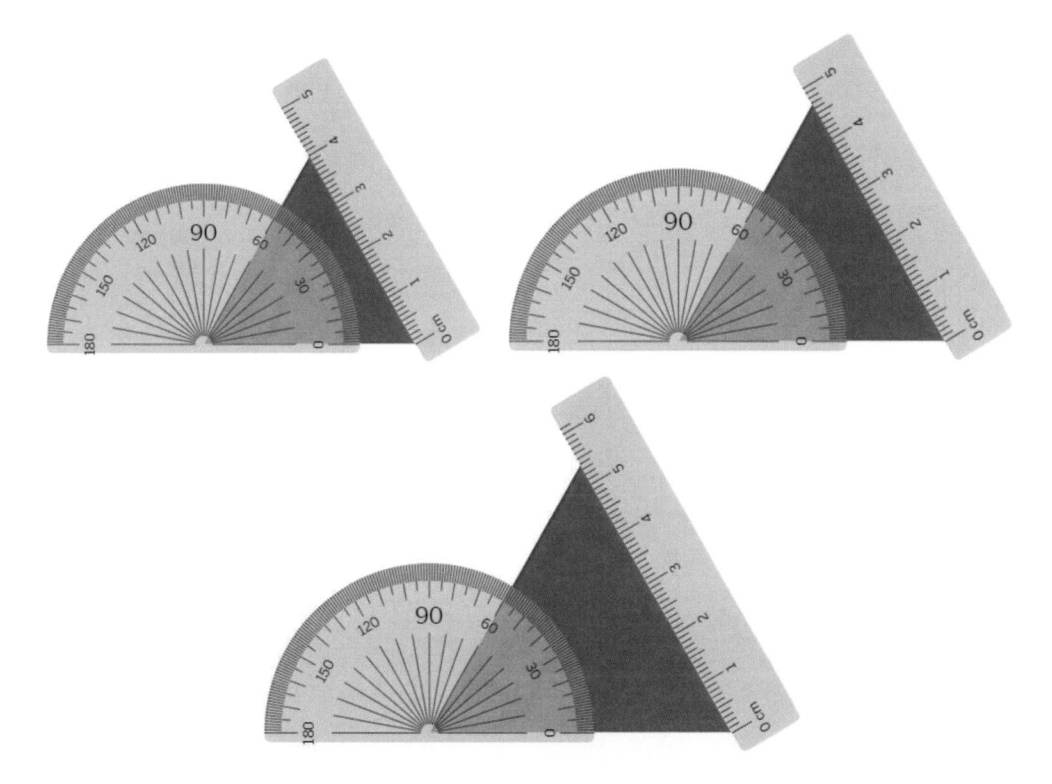

Encourage students to infer that the statement is never true for triangles with equal side lengths to have interior angles of different sizes as the equal side lengths force the interior angles to be equal in size too.

Notes: Triangles in the examples are from Polypad.org.

This activity can be done by cutting triangular shapes from paper and measuring their respective interior angles.

Students can further investigate properties of similar triangles and congruent triangles.

> **GS003 Statement:** A triangle has perpendicular sides.
>
> **Answer:** SOMETIMES TRUE
>
> **Manipulatives:** Geoboard

This statement can be shown as sometimes true through the use of a *Geoboard* with rubber band along with the knowledge of the term perpendicular. Isometric paper can be used as an alternative for a geoboard.

First let's consider creating the different types of right-angled triangles.

The diagrams below show the possible right-angled triangles that can be created. Discuss the impact of orientation with the students. Does the orientation imply a different type of triangle exists if the size is the same? This provides a window of opportunity to discuss the term congruence. Thinking should be in line with what congruence means to shape is the same as what equals means to number.

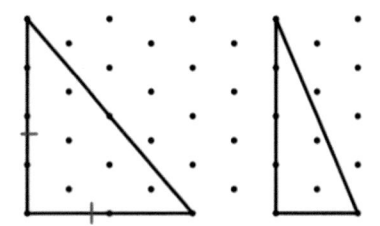

In the diagrams above, the dash on the sides represents equal side length. Discuss with students whether the length of the perpendicular sides has an impact on the angle created where they meet. Support articulating their findings with using vocabulary associated to the properties of the triangle.

Now let's consider other types of triangles. Some examples can be seen in the diagram below.

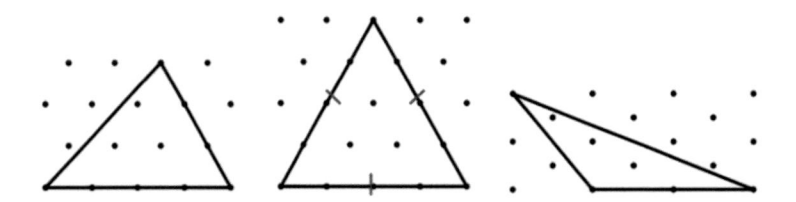

Here, students may use their understanding of a 3-sided polygon (a triangle) and declare that other triangles exist which do not have perpendicular sides. Thus revealing for these cases that the statement is not true.

From the above students can conclude that the statement is sometimes true.

Encourage students to notice any patterns through their creations, such as: a triangle can only have one obtuse angle; a triangle cannot have two pairs of perpendicular sides; a triangle with one pair of perpendicular sides can be an isosceles right-angled triangle; for a triangle to be isosceles, its pair of angles with equal value must be greater than 0° and less than 90°.

Notes: Triangles in the examples are from MathsBot.com.

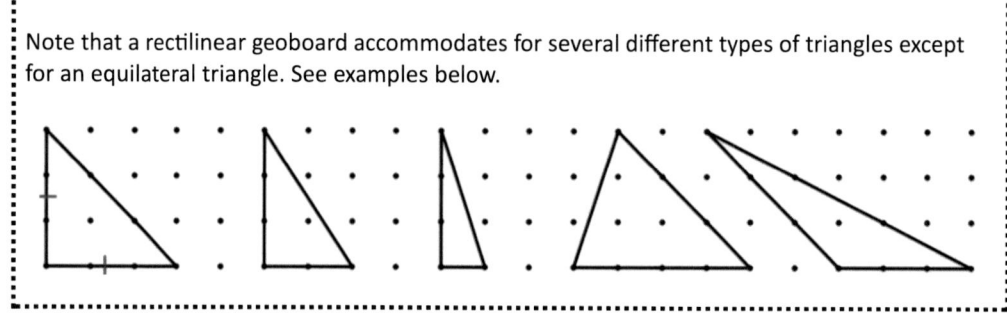

Note that a rectilinear geoboard accommodates for several different types of triangles except for an equilateral triangle. See examples below.

GS004 Statement: An equilateral triangle has one line of symmetry.

Answer: NEVER TRUE

Manipulatives: Equilateral Triangles

This statement can be shown as never true through the use of different sized *Equilateral Triangles*.

The approach used to investigate this statement is by applying the knowledge of symmetry and marking the interior angles of an equilateral triangle with different colours as seen in the diagram on the right.

If using paper equilateral triangles, then they can be folded by focusing on each different coloured vertex, as seen below.

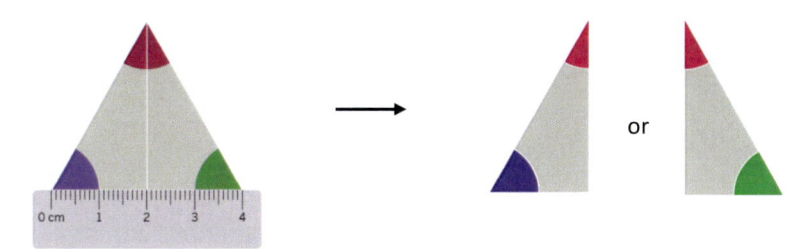

Irrespective of the fold, what should be noticeable is the angle that has been halved in size along with its corresponding side showing the line of symmetry at the fold. When the same strategy is employed with the other vertices, the diagrams below show the characteristic features of equal dimensions, again revealing the line of symmetry at the fold, shown with a dashed line.

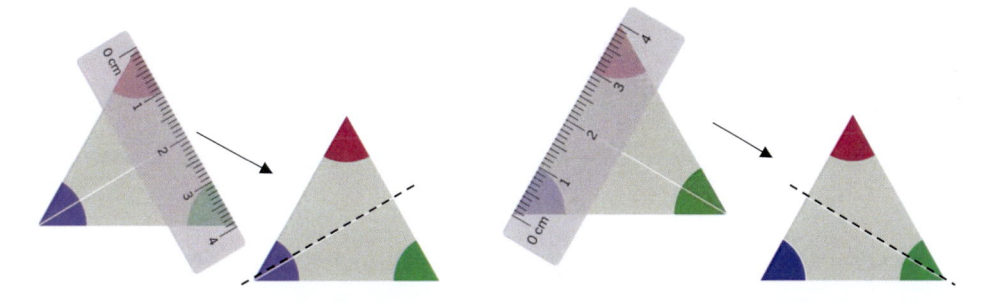

It can be seen from the above examples that the statement is never true as an equilateral triangle has 3 lines of symmetry. Students should investigate different sized equilateral triangles to conclude whether the statement is always, sometimes or never true.

Notes: Triangles in the examples are created from Polypad.org.

Further investigation about the relationship between a regular polygon and its lines of symmetry should reveal that the number of sides in a regular polygon equates to the number of lines of symmetry - refer to GS007.

> **GS005 Statement:** A square is a rectangle.
>
> **Answer:** ALWAYS TRUE
>
> **Manipulatives:** Rectangles

This statement can be shown as always true through the use of different sized *Rectangles* along with the definition of a rectangle and a square.

The approach used to investigate this statement is by adjusting the sides of a rectangle and applying the knowledge of the definition of a rectangle along with comparing it to that of a square.

The diagrams below show the checking of the dimensions of a 5cm by 3cm rectangle. The set square is used to check the sides are perpendicular, thus, confirming the interior angles to be right angles.

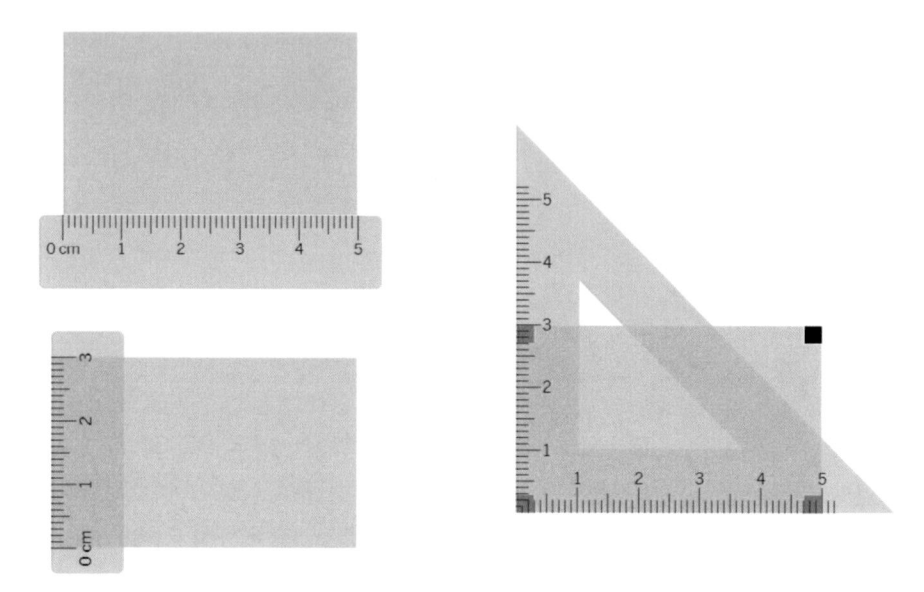

Keeping one of the sides a constant value and adjusting the length of its adjacent side provides an opportunity to identify patterns. The diagram below shows the labelling of the rectangles and for the purpose of the investigation, the length AD is kept constant at 3cm.

Students may jot their findings as shown in the table on the following page.

Shape	Side AD=Side BC	Side AB=Side DC	Shape Name
	3 cm	1 cm	Rectangle
	3 cm	2 cm	Rectangle
	3 cm	3 cm	Square
	3 cm	4 cm	Rectangle

From the above table, it can be seen that a square can be created from the properties of a rectangle and with further inspection, a square has properties that define it as a regular quadrilateral – a shape with all interior angles to be of equal value and all its side lengths to be of equal value. Therefore, the statement is always true.

Notes: Rectangles in the examples are created from Polypad.org.

Further investigations can be undertaken on other quadrilaterals with similar properties. For example: A square is a rhombus, A rhombus is a square, A rectangle is a parallelogram.

A square is a quadrilateral with four equal length sides where the opposite sides are parallel, four right angles, four lines of symmetry, rotation symmetry of order 4 and the diagonals bisect at right angles.

A rectangle is a quadrilateral with two pairs of equal length sides where the opposite sides are parallel, four right angles, two lines of symmetry, rotation symmetry of order 2.

> **GS006 Statement:** A kite is a rhombus.
>
> **Answer:** NEVER TRUE
>
> **Manipulatives:** Geoboard or Isometric Paper

This statement can be shown as never true through the use of a *Geoboard* with rubber band along with the definition of a kite and a rhombus.

The approach used to investigate this statement is by using an isometric geoboard with rubber bands. The kites are created by keeping one pair of equal length sides constant and adjusting the other pair of equal length sides incrementally as seen in the diagrams below.

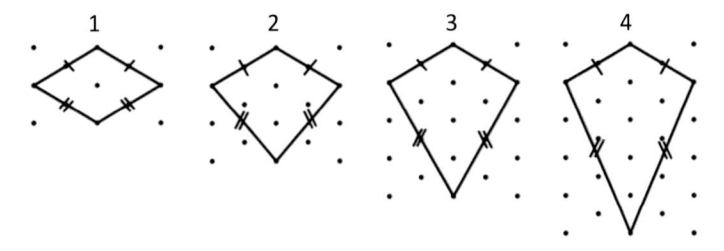

The equal length sides are identified by the single dash and double dash markings on the sides. Notice that the sides of shape 1 are all equal in length. Hence, creating a 4-sided shape with equal lengths. This shape is identified as a rhombus and not as a square.

Encourage students to articulate their findings using the properties of the kite and how they connect to those of a rhombus. Discussing the reverse statement – 'A rhombus is a kite' with students can further support their findings.

From the above, we can state that the statement is never true. Students can relate the connection to be similar to the relationship between a rectangle and a square - a square is a special kind of rectangle but, a rectangle is not a square.

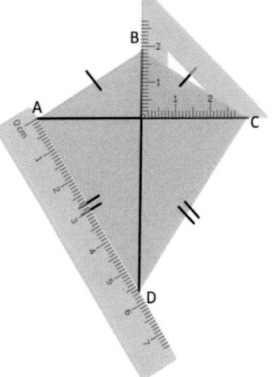

An example of how students may check their drawings is shown in the diagram on the right. Here the set square is used to check the diagonals intersect at 90° and the ruler to measure the side lengths.

> **Notes:** The shapes on the Geoboard are created from MathsBot.com and the kite with the measuring instruments is created from Polypad.org.
>
> A kite is a quadrilateral with two pairs of equal length sides and one pair of opposite equal angles (Angle BAD = Angle BCD in the diagram above). It has one line of symmetry, and the diagonals bisect at right angles.
> A rhombus is a quadrilateral with four equal length sides where the opposite sides are parallel, opposite angles are equal, two lines of symmetry, rotation symmetry of order 2 and the diagonals bisect at right angles.

> **GS007 Statement:** A square has 4 lines of symmetry.
>
> **Answer:** ALWAYS TRUE
>
> **Manipulatives:** Paper Squares

This statement can be shown as always true through the use of different sized *Paper squares* along with the definition of symmetry.

The approach used to investigate this statement is through folding a paper square and using the definition of line of symmetry. The line of symmetry in the diagrams below is shown using the dashed lines. The square is folded on that line showing the two halves are identical in shape and size. Therefore, confirming the fold to be the line of symmetry. The vertices have been labelled to provide clarity.

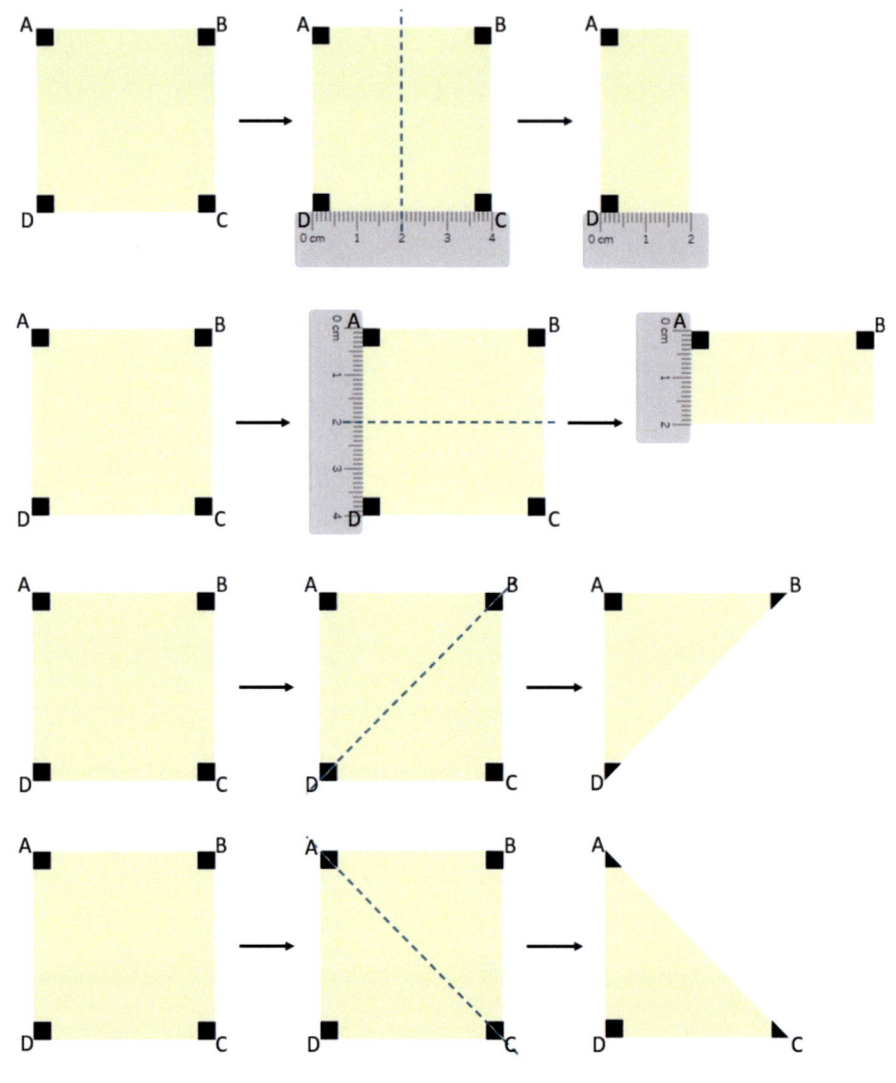

Encourage students to use different sized squares to support their generalisation about the number of lines of symmetry that exist in any given square.

The diagrams above show there are 4 lines of symmetry in a square, therefore the statement is always true.

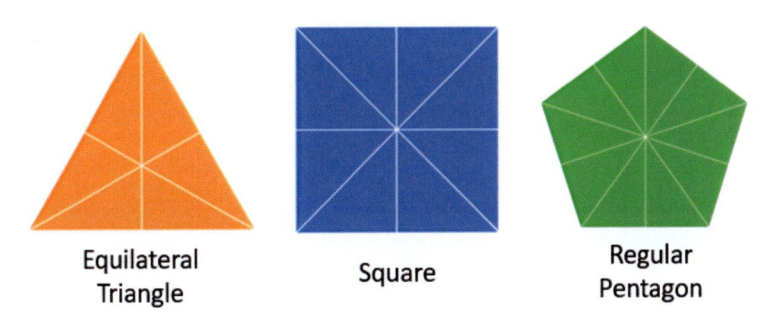

> **Notes:** Squares in the examples are created from Polypad.org.
>
> There are different ways to fold a square in half, but do they show the symmetry of the square? Hint: think of the area of half of the shape.
>
> Further investigations into the lines of symmetry of other quadrilaterals can be explored in a similar way.
> Also, investigation into the lines of symmetry of other regular polygons can be explored to generalise for all regular polygons – the number of lines of symmetry of any regular polygon is equal to the number of sides of that regular polygon. For example: an equilateral triangle has all 3 sides of equal length and has 3 lines of symmetry, a square has all 4 sides of equal length and has 4 lines of symmetry, a regular pentagon has all 5 sides of equal length and has 5 lines of symmetry as shown in the diagrams here.
>
> Equilateral Triangle Square Regular Pentagon

> **GS008 Statement:** A trapezium has two sides of equal length.
>
> **Answer:** SOMETIMES TRUE
>
> **Manipulatives:** Geoboard

This statement can be shown as sometimes true through the use of a *Geoboard* with rubber band along with the definition of a trapezium – a quadrilateral with one pair of parallel sides.

The above definition is used to explore and investigate the statement by constructing trapeziums under specific conditions.

First, let's consider the pair of parallel sides to be equal in length, identified by the arrow on the lines. In this case, when creating the trapezium on the geoboard, visualise the pair of parallel sides as equal in length a specified distance apart, as seen in the diagrams below. Then, complete creating the trapezium to ensure it is a closed shape with 4 sides. Through inspection of the shapes, it can be seen from the diagrams that the other sides are parallel as well, and equal in length too. Thus, the resulting shape is a rectangle/square/ parallelogram. As these have two pairs of parallel sides, they are not classified as trapeziums.

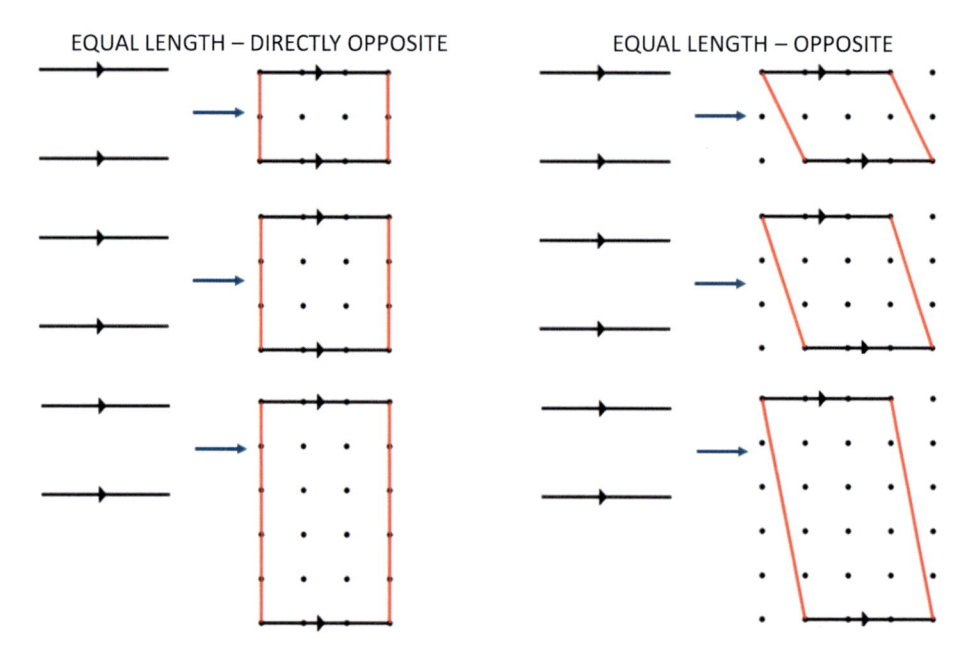

EQUAL LENGTH – DIRECTLY OPPOSITE   EQUAL LENGTH – OPPOSITE

Discuss with students the findings of the shapes that have been created as a result of the condition of one pair of parallel sides of equal length. Possible misconception as a result of the above could be that a trapezium is a rectangle/square/parallelogram. In which case, revisit the definition of a trapezium for clarity.

Now consider the parallel sides to be of unequal length. In this case, a systematic approach is used such that the length of only one parallel side is adjusted whilst the distance between the parallel sides is constant.

All three categories of trapezium should be explored. Students may choose to draw and identify the trapeziums as shown below. Encourage students to explore several within a category to support their findings and their generalisations.

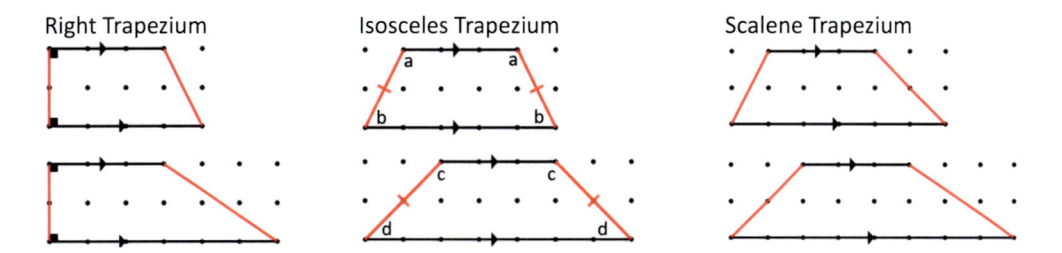

Right Trapezium   Isosceles Trapezium   Scalene Trapezium

From the above examples the statement appears to be sometimes true as the isosceles trapeziums have a pair of opposite sides of the same length.

Finally, the condition that has adjacent sides of equal length is explored. Here the sides of equal length are marked with a dash. Note the use of isometric geoboard/paper used to create shape 2 and 3.

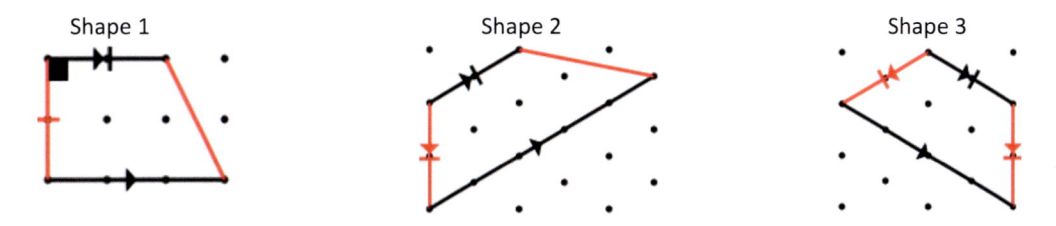

Shape 1   Shape 2   Shape 3

From the above explorations, the statement is sometimes true.

Discussions about the similarities and differences between trapeziums and other quadrilaterals with similar properties can support students in articulating their findings.

> **Notes:** Quadrilaterals in the examples are created from MathsBot.com.
>
> The North American and Canadian name for a trapezium is a trapezoid.

> **GS009 Statement:** Any quadrilateral can be made from two triangles.
>
> **Answer:** ALWAYS TRUE
>
> **Manipulatives:** Geoboard/Rectilinear or Isometric Paper

This statement can be shown as always true through the use of a *Geoboard* with a rubber band.

The assumption that is made to create the quadrilaterals is that the shape must have a common side from both the triangles. This means that triangles joining/touching at a vertex are not allowed. The diagrams below provide further clarity about this idea.

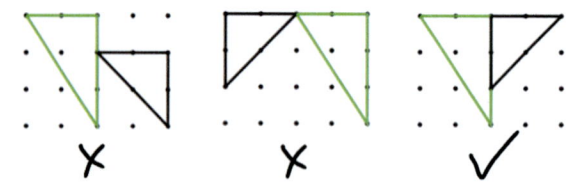

The ticked example above shows that an irregular quadrilateral can be formed using two triangles.

Now consider what happens with two **identical** scalene right-angled triangles. Notice the three different 4-sided shapes created as shown in the diagrams on the right. They are in different colours to emphasize the use of two triangles.
Encourage students to identify the quadrilateral that has been created and the reasons for their findings – How are they related to the properties of the quadrilateral?

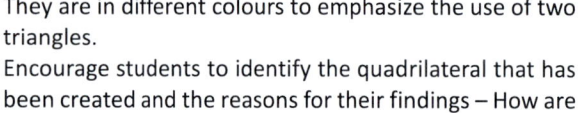

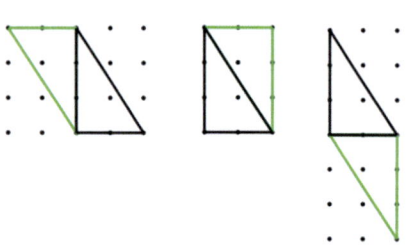

This suggests that a parallelogram and a rectangle can be created from 2 identical right-angled scalene triangles. Encourage students to predict which quadrilateral will be created from an isosceles right-angled triangle.

Now consider using the two **different** right-angled triangles. Example of using a scalene right-angled triangle with an isosceles right-angled triangle with one side of equal length can be seen here.

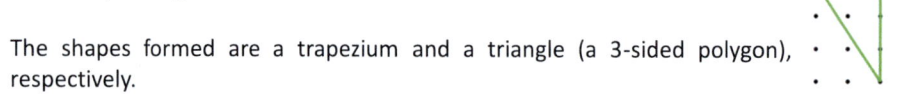

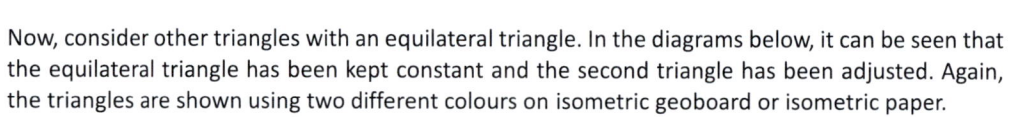

The shapes formed are a trapezium and a triangle (a 3-sided polygon), respectively.

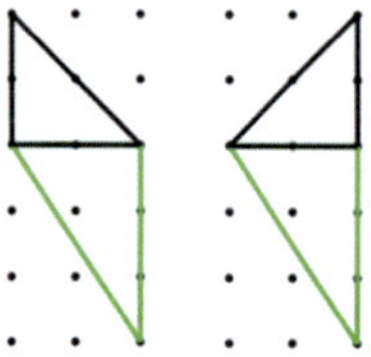

Now, consider other triangles with an equilateral triangle. In the diagrams below, it can be seen that the equilateral triangle has been kept constant and the second triangle has been adjusted. Again, the triangles are shown using two different colours on isometric geoboard or isometric paper.

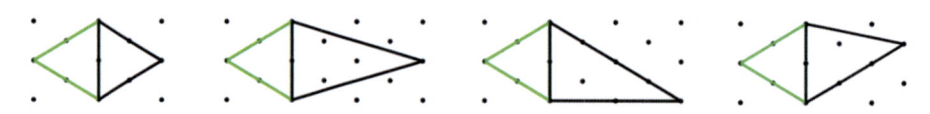

These suggest that a rhombus, a kite and a trapezium (two shown) can be created. Again, encourage students to identify the quadrilateral that has been created and the reasons for their findings – How are they related to the properties of the quadrilateral?

The above suggests that quadrilaterals can be created using two triangles. Therefore, the statement is always true.

Students might find during exploration that other polygons be formed too. Some examples are shown below using rectilinear and isometric geoboard/paper:

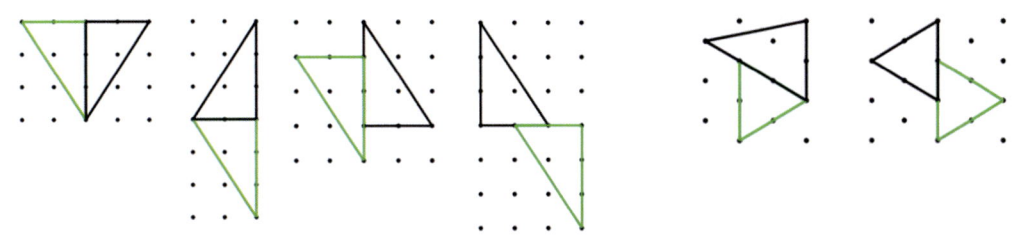

Notes: Polygons in the examples are created using Geoboard from MathsBot.com.

An alternative approach is to begin with a quadrilateral and separate it through one of its diagonals to form two triangles. This idea is shown in the examples below:

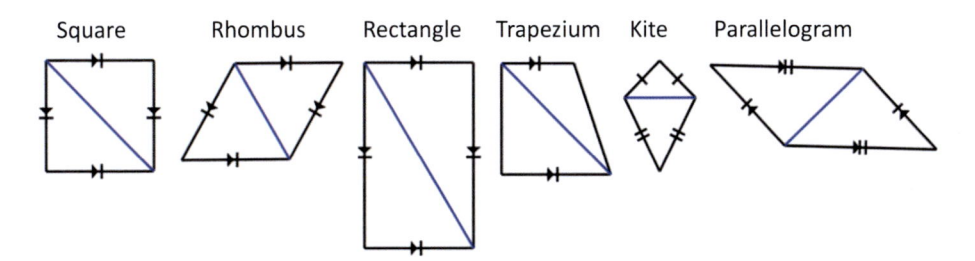

Square Rhombus Rectangle Trapezium Kite Parallelogram

Students can investigate this further by considering other polygons and their relationship with triangles. One such connection is that of the number of triangles constructed through the diagonals in any polygon is two less than the number of sides of that polygon. Refer to GS010.

Another exploration by students can be to find if a triangle can be formed from joining two triangles where one of them is an equilateral triangle.

> **GS010 Statement:** The interior angles of a pentagon total 360 degrees.
>
> **Answer:** NEVER TRUE
>
> **Manipulatives:** Pentagons

This statement can be shown as never true through the use of different sized and shaped *Pentagons* by using the knowledge of the sum of interior angles in any triangle.

The approach employed here is to create triangles inside the pentagon such that it looks like the pentagon is created from joining the triangles. The sum of interior angles of any triangle is equal to 180° is written inside each triangle to support any emerging patterns. See diagrams below.

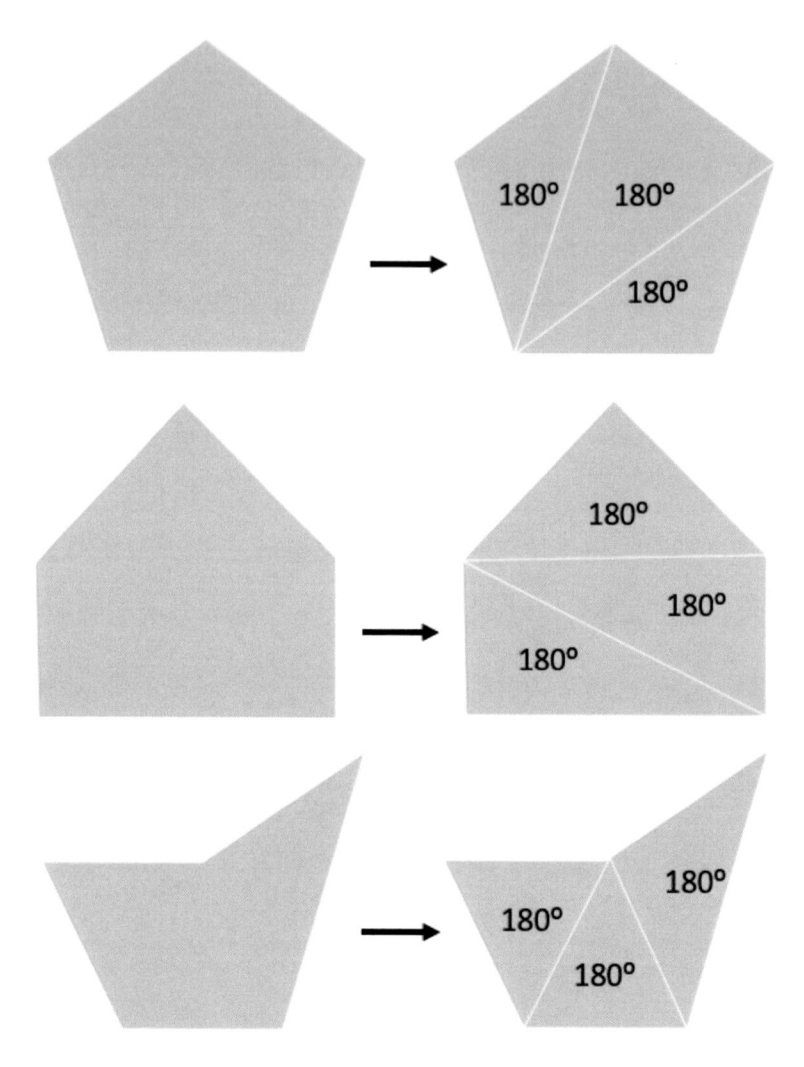

Encourage students to use several different sized and types of pentagons to support their findings.

As an alternative students can measure the interior angles and find the sum of the interior angles for every pentagon that is investigated. In this case, students should make jottings of the measured angles and their sum to support their findings.

From the diagrams above, it can be calculated that the sum of the interior angles of these pentagons is 3 x 180° = 540°. Therefore, leading one to conjecture that the interior angles of any pentagon cannot be equal to 360°. Hence, the statement to be never true.

Notes: The Pentagons were created using Polypad.org.

A further investigation may be to explore the statement 'A regular hexagon has interior angles of 120°'.

This can potentially lead to investigating the sum of interior angles of all polygons through triangles, leading to finding a rule to generalise for all polygons.

Rule: The sum of the interior angles of an n-sided polygon is equal to (n-2) x 180°. For example, The sum of the interior angles of a 5-sided polygon = (5-2) x 180° = 3 x 180° = 540°.

A similar approach is to use known facts of the sum of the interior angles in a quadrilateral as well. See diagrams below. These pentagons have been created using Geoboard from MathsBot.com.

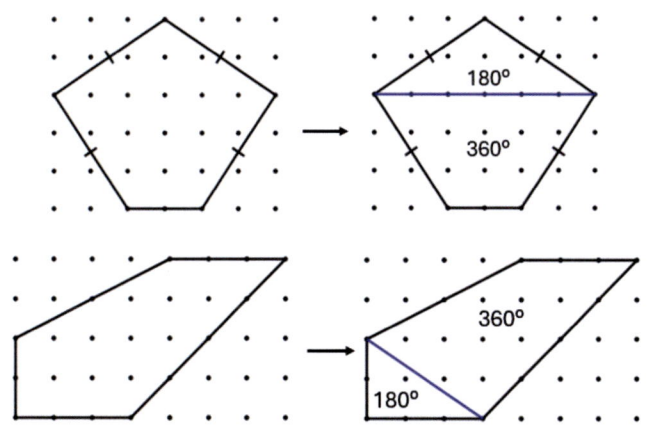

An alternative is to measure all five interior angles in a pentagon and calculate their sum.

GS011 Statement: A pentagon has two right angles.

Answer: SOMETIMES TRUE

Manipulatives: Geoboard

This statement can be shown as sometimes true through the use of a *Geoboard* with a rubber band by exploring different lengths of sides and angles.

Encourage students to begin by making some pentagons using geoboards. The geoboards printed here are from a screen format – but actual geoboards with pegs and bands are very satisfying to use if available.

Through exploration of a geoboard with pegs arranged in a rectilinear format the following pentagons can be made. Both these examples show two right-angles at the bottom two vertices leading us to see that the statement is true.

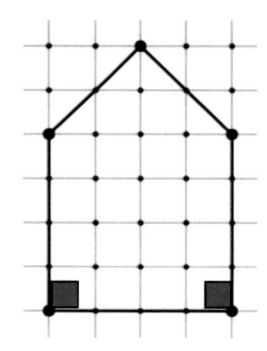

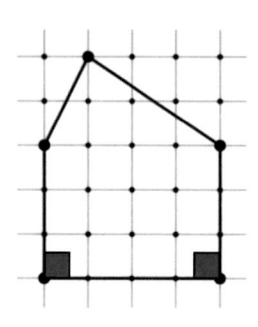

However, a pentagon doesn't have to have two right angles. See further examples below.

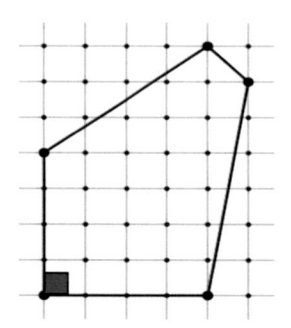

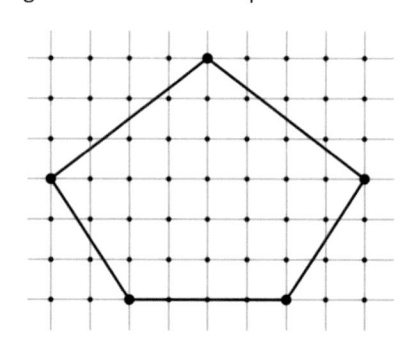

These show a pentagon with one right-angle and another with no right angles. This leads us to conclude that the statement is sometimes true.

If using geoboards with an isometric grid students could deduce some of the angles from knowing that an equilateral triangle has interior angles equal to 60°. A right angle can be made from one of the interior triangle angles and an edge that bisects one of them, thus the angle calculated as 60° + 30° = 90°.

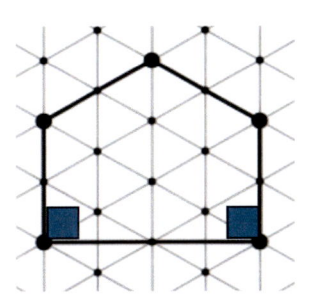

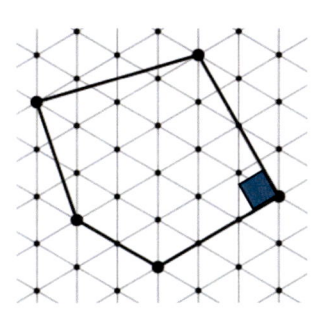

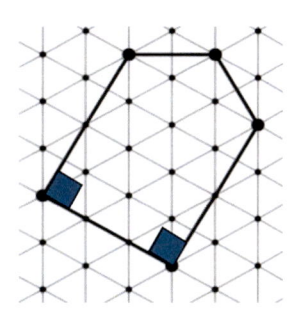

Encourage students to wonder about other properties: symmetry, hexagons, regular polygons and so on, and to explore their own statements with conditions made upon them.

> **Notes:** Geoboards used in the examples are from MathsBot.com.
>
> Geostrips can also be a useful manipulative (if available) to explore the properties of shapes. Right-angles can be shown by using a set square or something that clearly has right angles (such as the book of stamps as shown here).  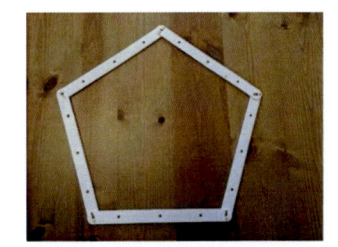
>
> Using paper with known angles, such as an A4 piece of paper which will have four right angles, students may also investigate what shapes can be made.
> The piece of paper on the right (below) has a corner cut off, producing a pentagon with three right angles.

GS012 Statement: Hexagons have sides that are equal in length.

Answer: SOMETIMES TRUE

Manipulatives: Geoboard

This statement can be shown as sometimes true through the use of a *Geoboard* with a rubber band by exploring different lengths of sides and angles.

Encourage students to begin by making some hexagons using geoboards. The geoboards printed here are from a screen format – but actual geoboards with pegs and bands are very satisfying to use if available.

Through exploration of a geoboard with pegs arranged in an isometric grid the following hexagons can be made. These examples show hexagons that have equal length sides leading us to see that the statement is true in these examples.

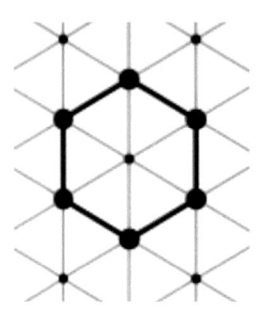

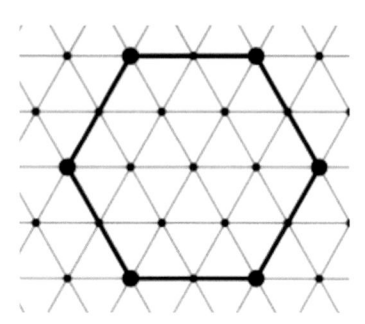

However, hexagons can have sides with different combinations of lengths. See examples below.

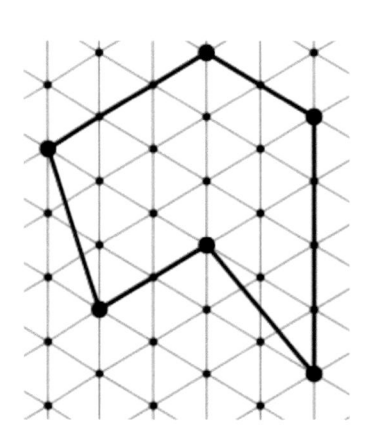

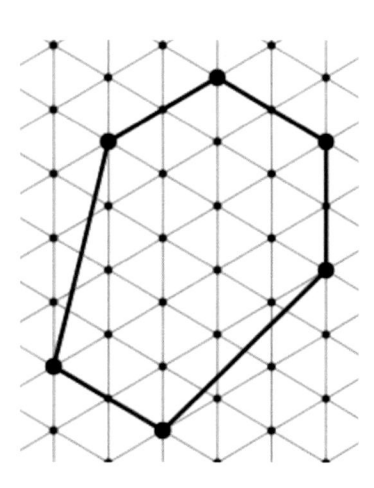

This leads us to conclude that the statement is sometimes true. Students should generalise that the statement will only be true if the angles are also equal and should discuss the differences between regular and irregular hexagons. Where students may talk about 'concave' and 'convex' shapes discuss the meaning of the words in relation to their hexagonal creations.

Notice that geoboards with a rectilinear grid do not create a regular hexagon even though it may appear to be one. Therefore, further discussions about how to identify the size of the side lengths will provide students with support in classifying regular and irregular shapes.

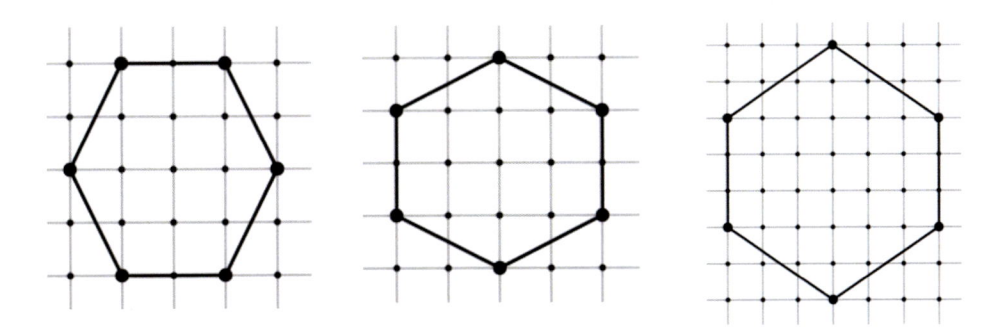

In the first two examples here, the lengths of the sides alternate between '2' and '$\sqrt{5}$'. Students would need to be familiar with Pythagoras to understand this, so careful measuring with a ruler may suffice for students who have not met this yet. It's a similar story with the third example, the sides here are '3' and '$\sqrt{13}$'.

Encourage students to wonder about other properties: symmetry, pentagons, other polygons and so on, and to explore their own statements with certain conditions made upon them.

Notes: Geoboards used in the examples are from MathsBot.com.

Geostrips can also be a useful manipulative (if available) to explore the properties of shapes.

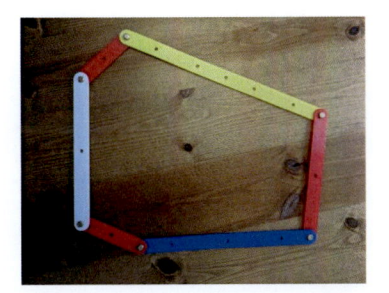

Paper can also be a useful manipulative. It is possible to fold a rectangular piece of paper into a regular hexagon shape. On the following page are examples of a regular hexagon (back and front) and an irregular hexagon.

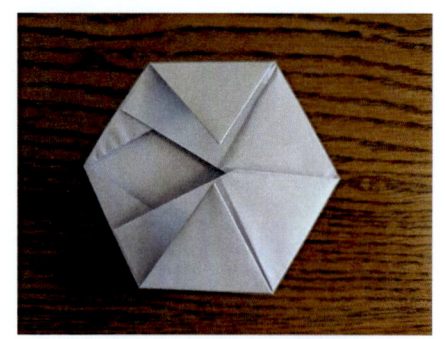

Regular hexagon - back

Regular hexagon – front

Irregular hexagon

GS013 Statement: The circumference of a circle is approximately three times the diameter.

Answer: ALWAYS TRUE

Manipulatives: Cylinders

This statement can be shown as always true through the use of a variety of *Cylinders* with some string or a tape measure.

Students will need to understand the words circumference and diameter in the statement to fully engage with the investigation. Encourage them to draw around the cylinder to create a circle and then to measure the diameter of their drawn circle. Also, encourage them to use string or a tape measure to measure the circumference. If using string, this can then be measured with a ruler to find its length. See photographs below of samples of cylindrical shapes and a way of measuring.

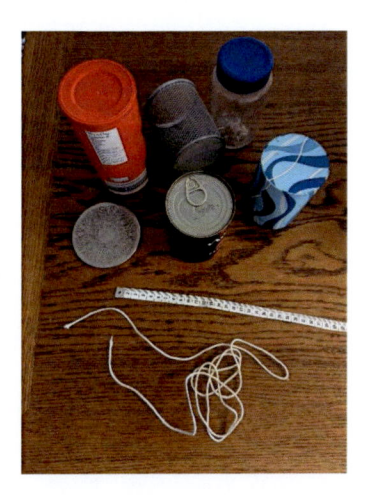

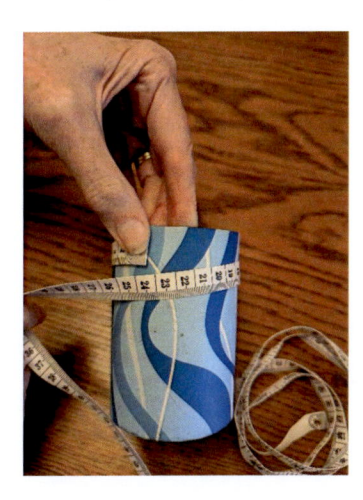

Encourage students to record their results in a systematic way, through recording the diameter and circumference in a table as suggested below.

Circle	Diameter	3 x Diameter	Circumference
A	7.5 cm	22.5 cm	23.8cm
B	8 cm	24 cm	25 cm

And so on…

Thinking…What is three times the diameter? How close is the measured circumference compared to the calculated value? Does the size of the cylinder have an impact?

Notes:

If appropriate, students could draw around their cylinders onto squared paper and estimate the diameter by counting the squares.

> **GS014 Statement:** If the area of a rectangle is 24 squared centimetres, then the side lengths are 4 cm and 6 cm.
>
> **Answer:** SOMETIMES TRUE
>
> **Manipulatives:** Square Counters or Maths Link Cubes

This statement can be shown as sometimes true through the use of *Square Counters* which could be maths link cubes.

Although, maths link cubes are a three-dimensional manipulative they are good for exploring a two-dimensional context such as this. As a result, a suggested approach is for the students to look at the top view of the arrangement created with the cubes to enable them to transfer into a two-dimensional pictorial representation as seen in the diagrams below. Square counters have been used to illustrate this idea.

First, let's consider using 24 counters (cubes) and explore what the lengths might be. Beginning with lengths of 6 and 4. See diagrams below. Colours have been used to support the recognition of rows and columns.

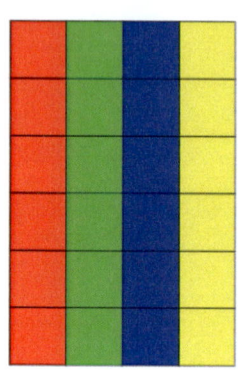

Clearly, the statement is true in this instance. Discuss with students whether the orientation of the rectangle is important or not.

Next, encourage students to manipulate and rearrange their counters (cubes) to make rectangles with different length sides. What else is possible while keeping 24 counters (cubes)?

This diagram below shows an arrangement where the sides are of lengths 2 and 12 but, the area remains as 24. Hence, this is an example where the statement is not true.

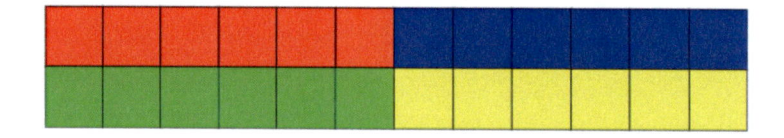

From the examples above the statement is therefore sometimes true.

Encourage students to find all possible rectangles with area 24 cm² and articulate the relationship between the area of the rectangle and its dimensions. Will a systematic approach provide all possible arrangements?

This could lead into an investigation into factors (see also MD001).

> **Notes:** Square Counters used in the examples are from MathsBot.com.
>
> Squared paper can be used to support students in this activity.
>
> Students should appreciate that multiplication is commutative and that an array of 6 x 4 will be equal to an array of 4 x 6. Hence the orientated versions are not included here.
>
> Other possible combinations are: 1 x 24, 2 x 12, 3 x 8, 4 x 6. Making four pairs of factors and eight factors in total.
>
> Students can further investigate the same relationship with other composite numbers, in particular square numbers (see MD022) and prime numbers.

> **GS015 Statement:** The perimeter of a rectangle is four times one of the sides.
>
> **Answer:** SOMETIMES TRUE
>
> **Manipulatives:** Square Counters or Maths Link Cubes

This statement can be shown as sometimes true through the use of *Square Counters* which could be maths link cubes.

Although, maths link cubes are a three-dimensional manipulative they are good for exploring a two-dimensional context such as this. As a result, a suggested approach is for the students to look at the top view of the arrangement created with the cubes to enable them to transfer into a two-dimensional pictorial representation as seen in the diagrams below. Square counters have been used to illustrate this idea.

First, let's consider a rectangle with lengths of 6cm and 4cm, as we need to start somewhere! This shows an area of 24 squared centimetres. The rectangle could be shown in different orientations as shown below.

Now, using the perimeter for a rectangle - the sum of the length of the four sides, for the example above: 4 cm + 6 cm + 6 cm + 4 cm = 24 cm.

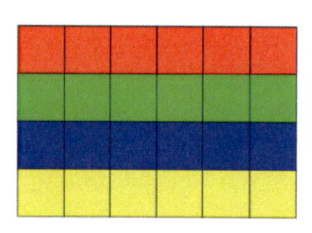

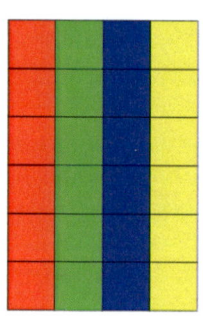

The statement states that the perimeter is four times one of the sides. Which side should be chosen? If length of 4 cm is selected, then the perimeter is 4 x 4 cm = 16 cm. If length of 6 cm is selected, then the perimeter is 4 x 6 cm = 24 cm. In one case, the statement is true and in the other it is not true.

Next, let's consider a different rectangle size, this time with lengths 2 cm and 12 cm – see below.

The perimeter is 2 cm + 12 cm + 2 cm + 12 cm = 28 cm.

Again, there are two different lengths of sides. Considering each of these with regards to the statement leads us to a perimeter of 4 x 2 cm = 8 cm and 4 x 12 cm = 48 cm, neither of which is the correct perimeter of 28 cm. Therefore, the statement is not true.

Encourage students to explore rectangles of varying sizes to support them in articulating their conjectures and generalisations.

Amongst their exploration, students may come across squares and decide not to investigate a square because of their understanding of a rectangle. In such a situation, discuss the connection between a square and a rectangle through their properties. This implies that if a rectangle whose sides are of equal length (a square), then the statement will be true. See diagrams below.

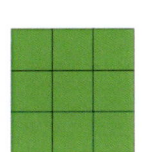

Perimeter is 3 + 3 + 3 + 3 = 12.
Length of sides is 3.
4 x 3 = 12

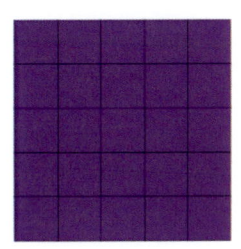

Perimeter is 5 + 5 + 5 + 5 = 20.
Length of sides is 5.
4 x 5 = 20

This leads us to an overall conclusion that the statement is sometimes true.

How many examples can students find where the statement is true? Is there a condition that supports their findings?

A parallel train of thought is to investigate the situation when the mean of two adjacent sides is considered instead. Is the perimeter four times the mean? Is this always, sometimes or never true?

Notes: Square Counters used in the examples are from MathsBot.com.

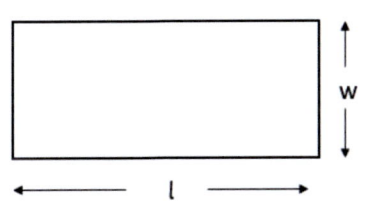

Students may generalise their findings using algebraic notation: P = 2(W+ L) (where P is perimeter, W is width and L is length).

If w = l as in the case with a square (a special kind of rectangle) then:
P = 2(W + L)
= 2 (W + W)
= 2(2W)
= 4W

> **GS016 Statement:** The area of a triangle is $\frac{1}{2}$ x base x height.
>
> **Answer:** ALWAYS TRUE
>
> **Manipulatives:** Squared Paper

This statement can be shown as always true through the use of *Squared paper* and the knowledge of area of a rectangle. There is an assumption that students should be familiar with the different types of triangles and that not all triangles have a right-angle.

However, let's begin by considering a right-angled triangle with base 6 cm and height 3 cm. Then create a rectangle with base 6cm and height 3cm around the triangle.

The right-angled triangle can be seen in the rectangle as though the rectangle has been split in half along one of its diagonals. This forms two congruent triangles. A check for congruency, by carefully placing one triangle on top of the other will show that they cover the same area. Hence showing that the area of the triangle is half of that of the rectangle.

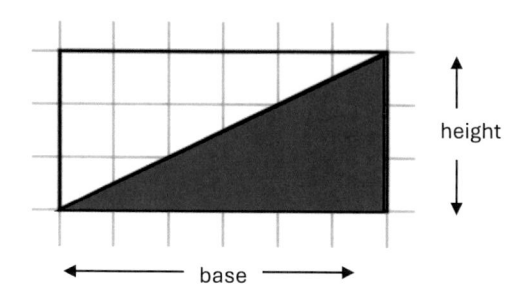

If the area of this rectangle is 6 cm x 3 cm = 18 cm², then the area of the triangle is $\frac{1}{2}$ x 18 cm² = 9 cm².

Encourage students to explore their findings through other sizes of right-angled triangles. From this, students might conjecture that the area of a triangle is $\frac{1}{2}$ x base x height.

This statement appears to be true for right-angled triangles.

Next, consider an isosceles triangle that does not have a right-angle. The diagram below shows a base of 4 cm and perpendicular height of 6 cm. Now let's create a rectangle around it.

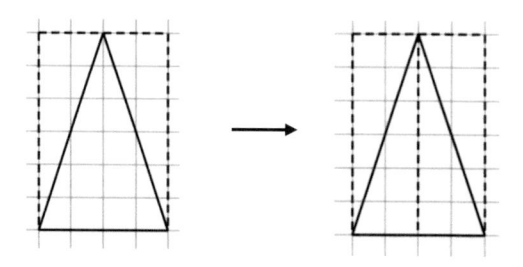

The outer rectangle can be split into smaller rectangles by drawing a perpendicular line through the top vertex of the triangle, creating two identical smaller rectangles.

The area of one of these smaller rectangles is 2 cm x 6 cm = 12 cm². Therefore, the area of the larger rectangle is 2 x 12 cm² = 24 cm². Students should predict that the area of the original triangle (left above) is 12 cm² (half of the area of the large rectangle).

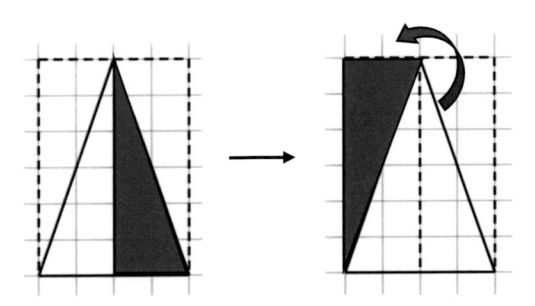

Again, students may wish to cut out their triangles and check that it will make one of the smaller rectangles. This results in the area of the smaller rectangle to be equivalent to the area of the original triangle. (Which is also half of the area of the larger rectangle.)

Encourage students to explore other isosceles triangles.

From this, the statement appears to be true for isosceles triangles.

Finally, let's consider a triangle where the perpendicular height of the triangle is not obvious.
In the example below, the triangle has a base of 4 cm.

By creating a right-angled triangle (shown on the right below), we can use what was conjectured above - area of a triangle is half x base x height. This is $\frac{1}{2}$ x 2 cm x 4 cm = 4 cm².

Similarly, the area of the larger right-angled triangle is $\frac{1}{2}$ x 6 cm x 4 cm = 12 cm².

Therefore, the area of the original triangle is 12 cm² - 4 cm² = 8 cm².

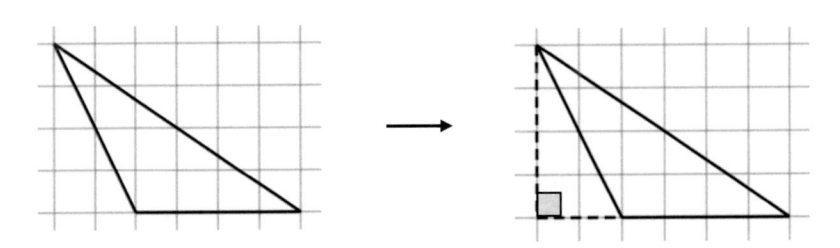

Now consider calculating the area of the original triangle without the use of the right-angled triangle. This means a perpendicular height needs to be considered as shown in the diagram on the following page.

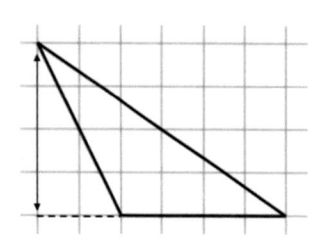

The area is $\frac{1}{2}$ x base x height.

In this example, the area is $\frac{1}{2}$ x 4 cm x 4 cm = 8 cm².

This shows that the height that must be used should be perpendicular to its base.

A common misconception is that one of the other sides is the height that is considered for the calculation - a slanting height should not be used!

> **Notes:** Geoboard from MathsBot.com is used in the examples as Squared Paper.
>
> Students need to appreciate that to find the area of a triangle, they will need the perpendicular height – which may need to be calculated in some way before they can use the formula:
> Area of a triangle = $\frac{1}{2}$ × b × h , where h = perpendicular height.

> **GS017 Statement:** Doubling the width of a rectangle will result in the area being doubled.
>
> **Answer:** ALWAYS TRUE
>
> **Manipulatives:** Squared Paper

This statement can be shown as always true through the use of *Squared paper* with rulers or through the use of *Geoboard* which was used to create these diagrams.

Ensure students can find the area of a rectangle and understand this to be length x width.

Let's begin by considering a rectangle that has a length of 2cm and a width of 1cm. Then, we can double the width making a width of 2.

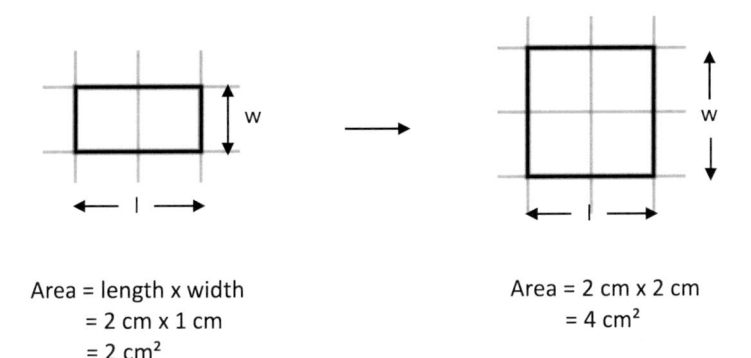

Area = length x width
= 2 cm x 1 cm
= 2 cm²

Area = 2 cm x 2 cm
= 4 cm²

The area of the double width rectangle is indeed double that of the initial rectangle. It's now 4 cm² rather than 2 cm².

Let's take this rectangle (square) and double the width again. See below.

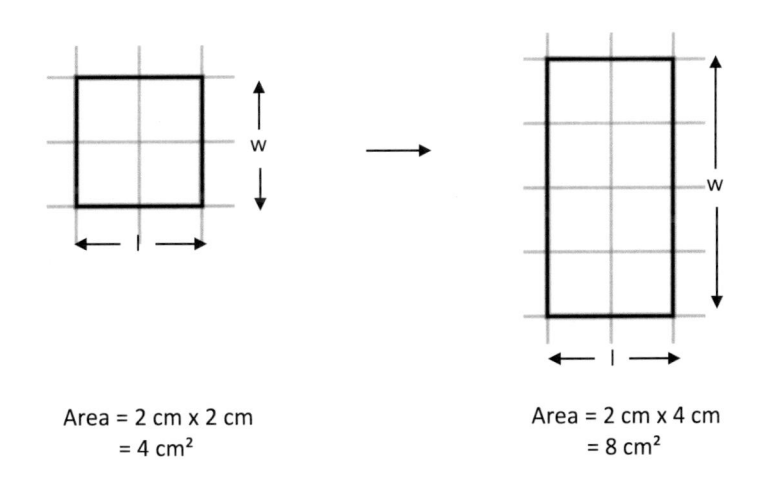

Area = 2 cm x 2 cm
= 4 cm²

Area = 2 cm x 4 cm
= 8 cm²

Again, this process doubles the area leading us to conclude that the statement is true.

Encourage students to continue working with integers and other types of numbers. Can negative integers be used? If not, why not? Can they explain their findings?

Notes: Geoboard from MathsBot.com is used in the examples as Squared Paper.

Students may like to adjust or add their own constraints to the statement and then investigate to see if they are always, sometimes or never true. For example:
- Doubling the length of a rectangle will result in the area being doubled.
- Doubling both the length and the width will result in the area being doubled.
- Doubling the length and halving the width will result in the area being doubled.

> **GS018 Statement:** Two rectangles have the same perimeter, so they will have the same area.
>
> **Answer:** NEVER TRUE
>
> **Manipulatives:** Squared Paper

This statement can be shown as never true through the use of *Squared paper* with rulers or through the use of *Geoboard* which was used to create these diagrams. There is an assumption that students are familiar with perimeter and area of rectangles.

Let's begin by considering a rectangle that has a perimeter of 24 cm. This number has been chosen as it has several factors which can be explored.

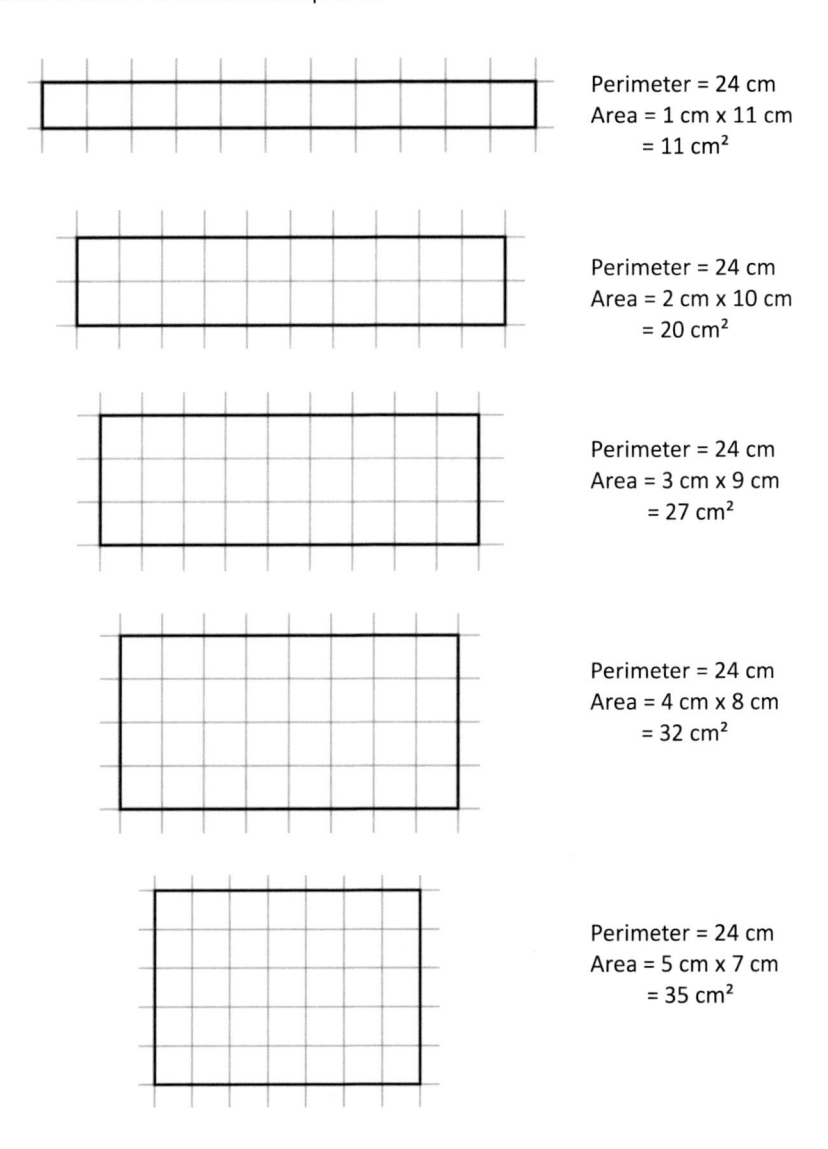

Perimeter = 24 cm
Area = 1 cm x 11 cm
= 11 cm²

Perimeter = 24 cm
Area = 2 cm x 10 cm
= 20 cm²

Perimeter = 24 cm
Area = 3 cm x 9 cm
= 27 cm²

Perimeter = 24 cm
Area = 4 cm x 8 cm
= 32 cm²

Perimeter = 24 cm
Area = 5 cm x 7 cm
= 35 cm²

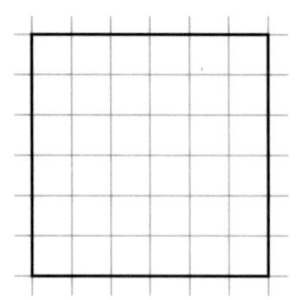

Perimeter = 24 cm
Area = 6 cm x 6 cm
= 36 cm²

This is a square – a special kind of rectangle.

These rectangles show that the perimeter remains at 24 cm, while the area changes each time. Leading us to believing that the statement is not true.

Where students have different orientations of the same rectangles, discuss the impact of this on the perimeter and area. Does it make a difference?

Encourage students to work with numerous examples. What do they notice? What do they wonder? What do they conclude?

Notes: Geoboard from MathsBot.com is used in the examples as Squared Paper.

Students may like to adjust or add their own constraints to the statement and then investigate to see if they are always, sometimes or never true:
- If two rectangles have the same area, then the perimeters will be the same.
- A rectangle can have an area that is numerically equivalent to its perimeter.
- A rectangle can have a perimeter that is half of its area.

GS019 Statement: A pyramid has an even number of faces.

Answer: SOMETIMES TRUE

Manipulatives: Pyramids

This statement can be shown as sometimes true through the use of *Pyramids*.

Let's consider the hexagonal pyramid below, is shown from several different aspects, including how it can be unfolded into a 6 "pointed star" net view.

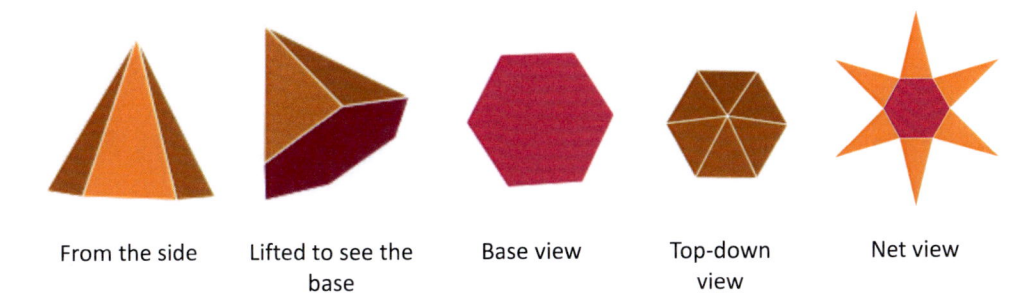

| From the side | Lifted to see the base | Base view | Top-down view | Net view |

This pyramid has 6 triangle faces meeting at the top (the apex) and one hexagon face on the base making a total of 7 faces. This is an odd number of faces, making the statement not true.

Students should continue exploring other pyramids with polygon bases. For example, if the pyramid has a square base, as in the famous ones in Egypt, then it would have 4 triangle faces and one square face making a total of 5 faces, also meaning that the statement is not true.

However, If the pyramid has a polygon face for the base with an odd number of sides, such as a triangle or a pentagon, then the total number of faces will be an even number. Triangle based pyramid: 3 triangle faces and the base making 4 faces in total. Pentagonal based pyramid: 5 triangle faces and the base making 6 faces in total.

This would mean that the statement is true. As there are examples for supporting both arguments, we can conclude that this statement is sometimes true.

Notes: 3D solids in the examples are from Polypad.org.

Students may like to adjust or add their own constraints to the statement and then investigate to see if they are always, sometimes or never true:
- A pyramid has the same number of faces as vertices.
- A pyramid has an odd number of edges.
- A pyramid has an even number of vertices.

Here are some examples of pyramids made from everyday packaging.

GS020 Statement: A prism has at least three rectangular faces.

Answer: ALWAYS TRUE

Manipulatives: Prisms

This statement can be shown as always true through the use of various *Prisms*.

Let's consider a hexagonal prism. Shown below are several different aspects, including how it can be unfolded into a net.

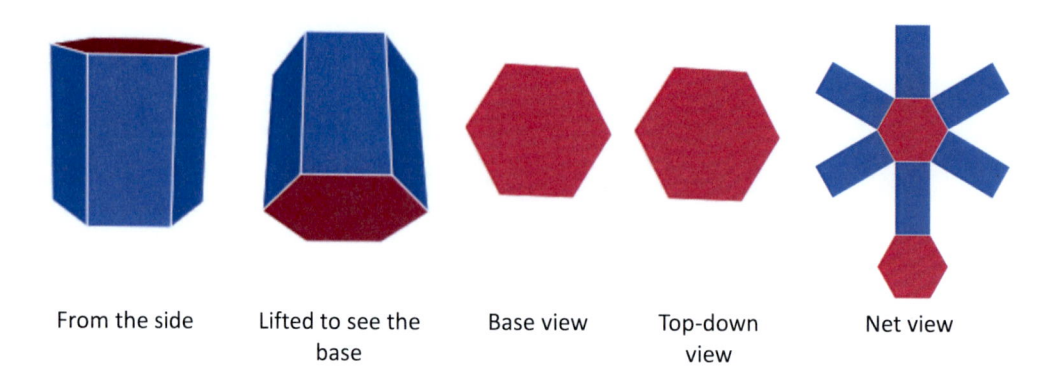

| From the side | Lifted to see the base | Base view | Top-down view | Net view |

This prism has 6 rectangular faces connecting the two hexagons on the opposite ends resulting in 8 faces in total and suggesting that the statement is true.

Students should investigate a number of different prisms and discuss their observations. They should be able to identify some of the common properties. A prism is a solid object with flat faces, congruent ends and the same cross section throughout its length, with the distinct feature that the cross section connects the polygons at its ends. A cube and a cuboid are the most common prisms found in the world around us. Is it possible to find an example without a rectangular face?

Students should articulate their observations through the use of shape related vocabulary. For example: If the prism had a square base, then it would have 4 rectangular faces connecting it to the square at the top of the prism. This prism is a 'cuboid', and the statement is true.
However, if the connecting faces were square, then we would have a 'cube'. And the statement would still be true as squares are special rectangles.

Students may tabulate their findings to support identifying patterns. For example:

Name of prism	Name and number of end Faces	Number of connecting rectangular faces	Total number of faces
Triangular Prism	Triangle - 2	3	5
Cuboid	Square – 2	4	6
Cube	Square - 2	4 square faces	6
Pentagonal Prism	Pentagon - 2	5	7

And so on…

Notes: Prism used in the example is from 3D Solids in Polypad.org.

A cylinder is not classified as a prism because its circular ends are not polygons.

Students may like to adjust or add their own constraints to the statement and then investigate to see if they are always, sometimes or never true:
- A prism has the same number of faces as vertices.
- A prism has an odd number of edges.
- A prism has an even number of vertices.

Finding pictures of examples of prisms in everyday contexts is helpful. Here are some we found:

> **GS021 Statement:** A regular polygon will tessellate with itself.
>
> **Answer:** SOMETIMES TRUE
>
> **Manipulatives:** Pattern Blocks

This statement can be shown as sometimes true through the use of *Pattern Blocks and other regular polygons.* Pattern blocks have a few regular polygons. However, to explore a wider range of polygons, students may also want to use paper polygons or stencils to supplement these.

An assumption is made that students know what tessellation means and are possibly aware of tessellations in the world around us, such as bathroom tiles, honeycomb, chessboard and so on.

Using the pattern blocks, students will see that the equilateral triangle, the square and the regular hexagon will tessellate – leading us to think that the statement is true. Can students explain why these shapes tessellate? What are the angles where the polygons meet?

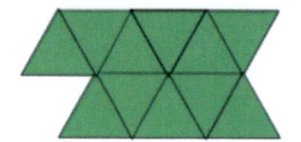

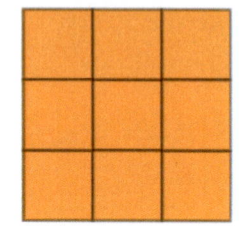

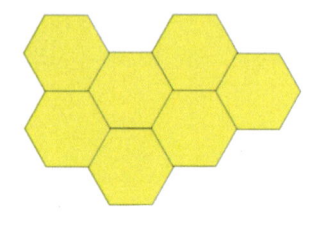

Equilateral Triangle　　　　　　　Square　　　　　　　Regular Hexagon

Whilst exploring with a regular pentagon students may notice a gap appears where they identify that another pentagon will not 'fit'. See example here: What could be predicted?

An example of further exploration with more pentagonal tiles is shown below.

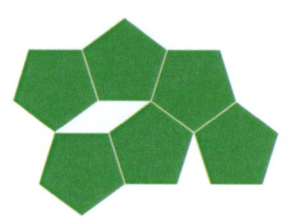

Can students work out why it does not tessellate?
Tessellations occur where the interior angles meet at 360° - visualising it as angles at a point. In the case of equilateral triangles, it can be seen that 6 triangles meet at a point which is 360°. Similarly,

it can be seen that four squares meet at a point, which is also 360°. However, in the case of the pentagon, the interior angles in a regular pentagon are 108° - is 360° a multiple of 108°?

This example leads to the conclusion that the statement is sometimes true.

> **Notes:** Pattern Blocks used in the examples are from MathsBot.com and the pentagon used in the examples is from Polypad.org.
>
> Students may like to adjust or add their own constraints to the statement and then investigate to see if they are always, sometimes or never true:
> - Quadrilaterals will tessellate.
> - A pentagon and a hexagon can tessellate together.
> - Regular octagons and squares can tessellate together.